GW00599314

RED SEA
SAFETY

To my parents

Contributing Photographers

F J Jackson
P J Vine
J E Randall
F Bavendam
R J Moore
Ron and Valerie Taylor
S J Hall
J Doran
A Fleetwood-Wilson

Illustrations

Jane Stark

RED SEA SAFETY

PETER VINE

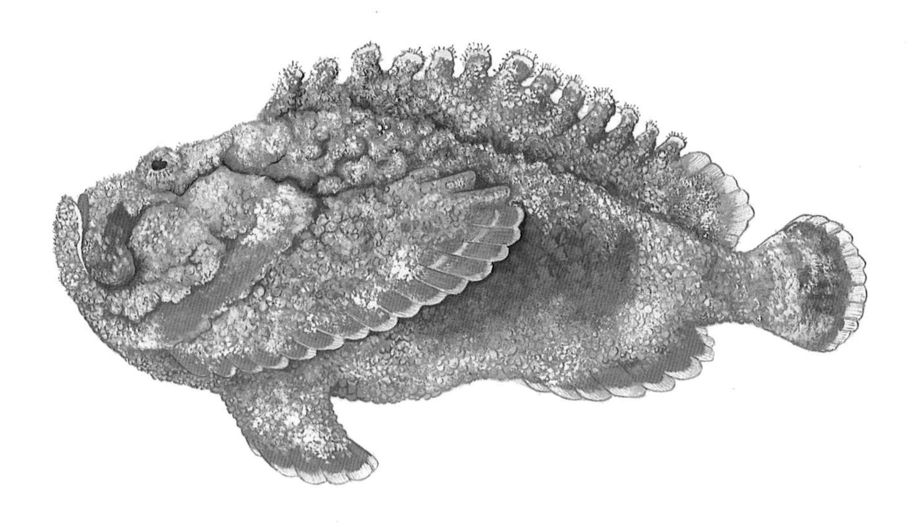

Guide to Dangerous Marine Animals

IMMEL
Publishing

Phototypeset in Gill Sans by
The Typesetting Company, London
Printed and bound in Japan by
Dai Nippon Printing Co, Tokyo
Design by
Patrick O'Callaghan Graphic Design

ISBN 0 907151 12 4

CONTENTS

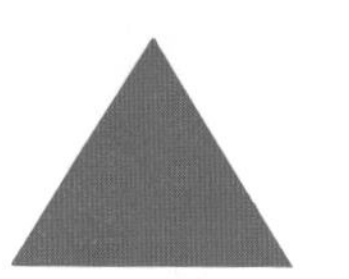

Almost as soon as the anchor was heaved overboard, myself and the six other divers hit the mirror calm sea. It was such a relief to cool-off after the scorching heat of that July afternoon. The old wooden Sambuk which had carried us to famous Shaab Rumi reef (once inhabited by Jacques Cousteau's aquanauts and their underwater village) had made a steady six knots all the way from the harbour, which we had left early that day. As the sun had risen higher in the sky we had sought relief from sun-burn by hiding in the shade of the furled sail, but for the last two hours of the trip there had been no escape from the blistering heat and hence our rapid plunge into the crystal clear cool waters of the Red Sea. It was wonderful to be underwater again and able to glide down to the reef terrace to experience the tranquility of this tropical aquatic paradise. This was, as my diving colleague, Douglas Allen, described it – "a magic place".

When Cousteau's team vacated the reef in 1965, they thoughtfully left behind the underwater garage which had been used for parking and repair of their diving vehicle. Now the domed yellow garage, which sits close to a steeply inclined reef-face, had a rich covering of live corals and was inhabited by fish instead of divers. After a brief stop to review our acquaintance with the garage and its residents, Doug and I headed on down along the reef slope towards the dark blue waters which had been such a happy hunting ground for Cousteau's divers. It was here that they had filmed sharks and manta rays cruising along the reef-edge and our thoughts were on them as we descended past huge gorgonian fans and luxuriant coral colonies, through an abundance of marine-life which we had rarely encountered.

:ousteau's Garage
: Shaab Rumi Reef
'hoto S J Hall

At around 100 feet deep we caught sight of a large shark swimming along the "drop-off" and we glanced around for a protected cranny where we could sit quietly and observe the comings and goings of this marine metropolis. Doug spotted it first and it took me a few seconds to recognise the object which he was gesticulating towards. It appeared to be a metal grill lying against the reef together with some sort of rope or cable. Whatever it was, it was man-made, and must have been part of the debris left behind by the Conshelf team. When we swam closer I realised why Doug had been so interested. It was in fact an old shark-cage which must have provided a safe haven for divers when they were photographing baited sharks. It had already been tied to the reef for ten years and there was little wonder that it was heavily encrusted with a "fuzz" of hydroids and other fauna.

In the years since Cousteau's early underwater

Grey Reef Shark
Carcharhinus amblyrhynchos
Photo Jim Doran

experiments, divers such as ourselves had become more familiar with Red Sea sharks and perhaps less cautious. We had almost forgotten in fact that this was "Highway 1" for what Doug referred to as "the biggies"! Glancing around to check on our own position I was greeted by a fairly close view of a medium sized "Silvertip Reefshark" (*Carcharhinus albimarginatus*) which looked as interested in us as we were in the old shark protection cage. It seemed like a good idea to try out the cage so we prised open the door and squeezed into the confined quarters surrounded by narrowly spaced metal bars. For a brief moment we felt safe and then we both realised our mistake. The inquisitive shark was not our problem – in fact we could not leave the cage fast enough! We were being stung so violently that I would have defied anyone to survive in that situation for more than about a minute. Whatever it was that had "got at us" seemed every bit as dangerous as the shark which remained in our vicinity. We had no choice but to take our chances with the sharks and escape from our hidden tormentors. Already I could see weal marks on Doug's arms and legs, so taking one look at each other, we headed back towards the surface. We both realised that we had foolishly brushed up against the bars of the shark-cage and were suffering the painful consequences of an unwelcome meeting with a stinging hydroid – probably, I told myself, *Lytocarpus philippinus*, but leaving positive identification for a future occasion, we spent the next hour on board the Sambuk, applying surgical spirits to our stings and thus immobilising any adhered stinging nematocytes.

On reflection later I realised that the so called dangers of diving or swimming on coral-reefs more often emanate from small and fairly inconsequential looking creatures than from the much maligned "dangerous animals" such as sharks and barracudas. Some of the worst mishaps associated with watersports arise from Man's foolhardy or ignorant actions in an environment where – however hard he tries to prove otherwise – he will never be anything other than a temporary guest.

In this book I have set out to provide a basis for understanding the potential dangers which may be encountered while diving among Red Sea coral reefs. Informed divers or reef-walkers may in fact do so in the knowledge that their chosen leisure activity is not inherently more dangerous than any other outdoor pursuit. Marine creatures which one encounters in the Red Sea are *not* out to attack Man, but may, under certain conditions of provocation, unleash their defence mechanisms and can thus unwittingly cause mild pain and temporary discomfort or more severe suffering and, in a few cases, even death of the unsuspecting and generally speaking unprepared visitors to their marine world.

As the saying goes: "there's more than one way to kill a cat". Marine creatures can inflict pain in a number of different ways and I have taken this broadly functional approach in

Stinging Hydroid
Lytocarpus phillipinus
Photo P J Vine

discussing the dangers associated with Red Sea coral reefs rather than adopting a format based on scientific classification of the organisms concerned and their individual characteristics. In many cases an animal which may be quite safe to hold in the hand is poisonous to eat, and when it comes to those creatures which can inflict painful or even lethal wounds, the rogues' gallery contains more than sharks and barracudas.

Coral growth on the Shab Rumi Reef Sudanese Red Sea Photo F J Jackson

However careful the reader may be while wading, snorkelling or SCUBA diving on Red Sea reefs, it is quite probable that sooner or later he or she will be faced with one of the dangers described in this book. Such an encounter may be a personal problem or that of a friend. Whichever occurs, it is important that rapid and informed action is taken to control the situation. Medical opinion varies in some cases about the best treatment for poisoning, stings or bites inflicted by various marine organisms but there is now sufficient experience of most of the accidents which tend to occur in the Red Sea to enable the recommendation of effective first aid. This book presents the latest available information on this subject and it should provide a useful handbook for all those who, like the author, enjoy spending their time in and around the Red Sea.

The book has been divided into a main text and a series of appendices. The latter are not directed at the general reader but are designed to provide medical practitioners with basic guidelines regarding specific aspects of marine bio-medicine. While every care has been taken to ensure that this information is correct and up to date it must be stressed that it is not intended to replace the experience and judgement of medics called to attend marine related cases. Refinement of treatment procedures in such cases results from communication of actual experiences by doctors working in the region. The author would be most interested to hear the views and comments of such medical personnel in order that future editions can be updated and treatment procedures improved.

Most Red Sea fish are perfectly safe to eat and the incidence of confirmed fish poisoning due to icthyotoxins is rare. There are nevertheless a few do's and don'ts to follow when buying, preparing or eating Red Sea foods and there are one or two rules which should always be followed. Despite this it would be possible for the uninformed consumer to eat Red Sea fish on every day of his or her life without ever experiencing the debilitating consequences of tropical marine fish poisoning. In the relatively rare cases however, where people do suffer from one of the recognised fish toxins, the effects are so dramatic and traumatic that one is unlikely to take unnecessary chances every again.

Research on tropical fish poisons has gone on for several hundred years. Knowledge of toxic Red Sea fish dates back long before this, to the time of the Ancient Egyptians who demonstrated their experience of the powerful toxicity of Puffer Fish by engraving the locally common species, *Arothron stellatus*, on several of their tombs. These date back to as long ago as 2,700 B.C., during the Fifth Dynasty and the tomb of Ti. Other similar engravings may be seen at the tombs of Méra, Giza, Ptah-hotep and Deir-el-Gebrawi. It is clear that these early civilisations had learned of the dire consequences of eating certain species of Red Sea fish and stories were passed from generation to generation until a general rule emerged. This was stated around 1,450 B.C. and is recorded in the Old Testament of the Bible:

(Deuteronomy 14, verses 9-10) where it is written: "of all

that are in the waters you may eat these: Whatsoever has fins and scales you may eat. And whatsoever does not have fins and scales you shall not eat; it is unclean for you".

Recent research on icthyotoxins among those fish species which occur in the Red Sea has underlined the validity of this ancient teaching. Among locally occurring species, the two fish families most likely to cause severe poisoning in Man are Puffer Fish (Tetraodontidae) and Moray eels (Muraenidae). Both these are effectively eliminated from the diet by the above ancient teaching which is almost three and a half thousand years old!

Most modern research on tropical fish poisons has taken place in the Pacific region but many of the implicated species are ones which also occur in the Red Sea and there is no reason to believe that similar toxicity does not occur in Red Sea fish. Indeed, in several cases this has been confirmed by discussions with local fishermen who are aware of the dangers of eating potentially toxic species. The most widespread form of fish poisoning associated with coral-reef fishes is known as Ciguatera. The name is derived from Cuba where residents became ill from eating turban shells locally called "cigua". It is now used as a general term to describe tropical marine fish poisoning in which a characteristic series of symptoms are caused by ingestion of a toxin referred to as ciguatoxin. It is not the only form of fish poison associated with coral-reef fishes although the term is often incorrectly used to describe other forms of fish poisoning. The situation is complicated by the fact that fish which are reported to be ciguateric may contain several toxins of different chemical composition and producing different symptoms.

What causes normally edible fish to become toxic? The question has been studied by a number of eminent scientists who have proposed several theories to explain the phenomenon. It has now been proved that the food chain is involved and that the ciguatoxin is concentrated in the livers of infected species. Many herbivorous fish such as members of the Surgeonfish (Acanthuridae) family have been shown to contain ciguatoxin and Red Sea species such as *Acanthurus sohal* and *Ctenochaetus striatus* are thus implicated. Toxic concentrations in such species are however much less than in certain predatory species which feed upon reef herbivores. Some of the highest levels of toxicity have been found in the Red Sea snapper *Lutjanus bohar* and in the Giant Moray, *Gymnothorax javanicus*.

A recent study of this subject carried out by Dr. J.E.

Giant Moray
Gymnothorax javanicus
Photo F Bavendam

Randall at Enewetak in the Marshall Islands (Randall, 1980) investigated 551 fish belonging to 48 species from Enewetak together with a further 256 fish (23 species) from Bikini and 12 large individuals of *Lutjanus bohar* from Rongelap. The method employed for testing toxicity levels was to feed the liver and viscera to laboratory held mongoose and to observe their responses. In Randall's study eight species of fish showed toxicity levels which could cause severe illness or death. These were *Gymnothorax javanicus* (Moray eel); the groupers: *Cephalopholis argus*, *Epinephelus hoedtii*, *E. microdon* and *Plectropomus leopardus*; and the snappers: *Aprian virescens*, *Lutjanus bohar* and *Lethrinus kallopterus*. Randall comments that had more specimens of the barracuda *Sphyraena barracuda*, the Jack *Caranx ignobilis* or the wrasse *Cheilinus undulatus* been studied, then these would also have been included in the above list.

Dr. Randall's examination of the feeding habits of ciguateric fish led him to draw a number of interesting conclusions. He stated that those carnivores which prey heavily on reef fishes are the most prone to be poisonous while those which eat mainly benthic crustaceans are least likely to be so affected. At the same time he recognised that several species which feed on molluscs or echinoids may become extremely toxic.

In an earlier study on this subject, Randall (1958) suggested that ciguatera tends to break out in areas of the reef where some form of ecological disturbance has recently taken place. He conjectured that the chain reaction may begin with those species of blue-green algae which are first to settle upon newly exposed surfaces such as ship wrecks, recently killed coral, or rubble associated with mining and dredging operations. This theory has been discussed by Prof. Banner from the University of Hawaii who stated that it is difficult to prove or disprove since there are cases which support the idea and others where outbreaks do not appear to coincide with such disturbances. In a subsequent study Randall pointed out that many of the fish which harbour significant concentrations of ciguatoxin are species high in the food chain which tend to have a wide swimming range and may therefore be caught at some distance from where their feeding gave rise to ciguatoxin build-up. Thus, he suggests that direct correlation between cause and effect in the case of ecological disturbances and ciguatera outbreaks may be complicated by migrations of the main species concerned in ciguatera poisoning. After a great deal of research into the nature and possible causes of ciguatera, the causative agent was finally revealed in 1979. Dr. Randall's earlier suggestion that blue-green algae were somehow involved turned out to be very close to the mark for the offending organism was identified as a dinoflagellate which is found living on blue-green algae. The organism has a mouthful of a name: *Gambierdiscus toxicus*. The proposal that localised environmental upheavals caused by natural or human agents may create conditions for increased growth of blue-green algae and associated population explosions of the offending dinoflagellate – thus leading to outbreaks of fish poisoning – now seems to provide a logical explanation for many recorded outbreaks of ciguatera. At the time of writing however it remains uncertain whether this particular dinoflagellate occurs in the Red Sea or whether other species, acting in a similar way, are involved in the various reported incidents of ciguatera type fish poisoning which have occurred in the area.

Toxins other than ciguatoxin may accumulate in coral-reef fishes. Banner (1967) reported on three different fish toxins which created differing effects in those who suffered. He showed that ingestion of toxic livers from the Moray eel *Gymnothorax*, snappers of *Lutjanus* and groupers belonging to *Epinephelus* could cause death by respiratory failure and they showed high anticholinesterase activity. A second group of fishes which included the giant wrasse *Cheilinus* and the surgeonfish *Acanthurus* could cause death by cardiac failure and toxin from these fish had a lower anticholinesterase activity. A third group included the Barracuda (*Sphyraena*)

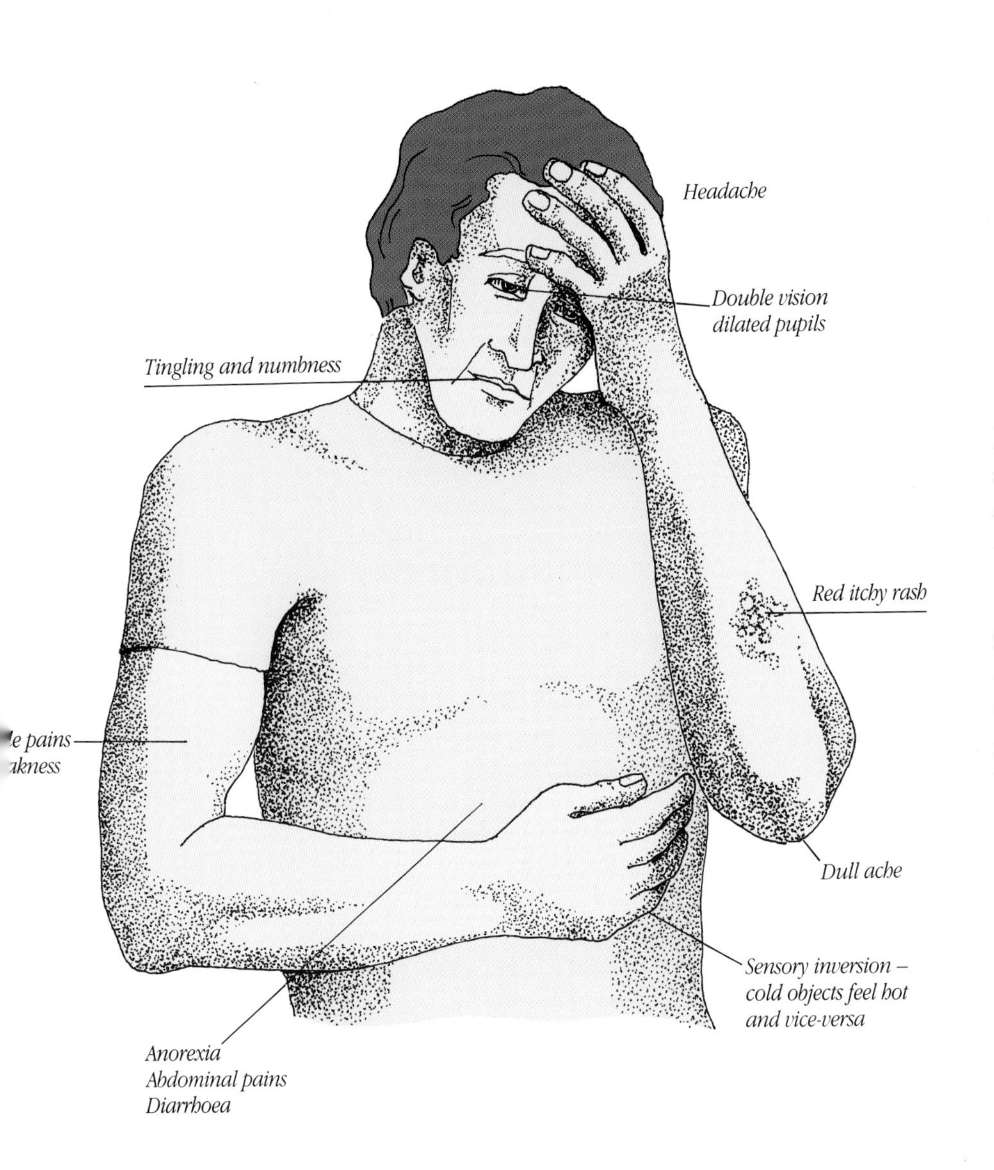

FIGURE 1

General symptoms of ciguatera – fish poisoning.

and Jacks (*Caranx*) and it was intermediary between these first two groups.

The poison known as ciguatoxin has been isolated by various techniques and Professor Banner from the University of Hawaii stated that the process of extraction and purification of it went through 14 modifications in as many years. The latest system he described gave 10ppm of highly toxic raw fish – and the purification yielded 5-10mg/kg and a toxicity of 0.025mg/kg when injected intraperitoneally into mice. Its chemical formula would appear to be $C_{35}H_{65}NO_8$ (based upon combustion).

The pharmacological action of ciguatoxin has been studied in detail. One of the major assumptions has been that it behaves as an anticholinesterase but this theory was thrown into doubt by Okihiro who reported on an attempt to treat a near fatal case of ciguatera with a cholinesterase regenerator (2-PAM). He stated: "The results were startling, unexpected and almost disastrous". The patient was just saved as a result of a bedside tracheotomy together with massive doses of atropine and magnesium sulphate! It was later shown that while there is some evidence of inhibition of cholinesterase in vitro, it definitely does not work this way in living systems. Hence the near fatal mistake!

It now appears that ciguatoxin increases the permeability of excitable membranes to Na+ ions. Professor Banner has commented that "The more acute symptoms associated with ciguatera poisoning appear to be consistent with the proposed mechanism of action and a rational therapy would be to treat the membrane effects of ciguatoxin with magnesium sulphate and calcium glutonate".

He adds that atropine has been successfully used to relieve symptoms associated with the disfunction of the autonomic nervous system. There may however be further mechanisms involved and the treatment may in future be revised based upon improved understanding of these. The above somewhat scientific account of ciguatera and the effects of ciguatoxin may tend to mask the really dramatic nature of this form of fish poisoning. As I have alluded above, the primary symptoms are neurological but in addition the sufferer may complain of severe stomach cramps, nausea and diarrhoea. A characteristic symptom which nearly always occurs is that of tingling lips, mouth and tongue. In some cases large areas of skin become itchy. General weakness, exhaustion and various aches and pains follow and partial or complete paralysis may occur. Another particularly characteristic symptom is "sensory inversion" in which hot objects feel cold. The victim often sweats profusely and may show a loss of reflexes. In acute cases the patient may become comatose and eventually die as a result of respiratory failure. The main symptoms usually occur within six hours of eating a toxic fish and full recovery may take months.

There are a number of preventive measures which can be taken to avoid the danger of ciguatera poisoning. In general it is the largest specimens of the carrier species which are most likely to be heavily infected with ciguatoxin. It is therefore a good rule to avoid buying or eating these big fish. Secondly, since the toxin accumulates in the liver and viscera it is best to clean fish thoroughly and to avoid fish livers. Thirdly, in situations where fish-poisoning has been known to occur or where one has reason to suspect ciguatera or any other form of fish poisoning, a local cat should be given some viscera from the fish before cooking. Such a test animal should then be observed for at least a day before cooking the fish for human consumption. An early sign of intoxication in cats is loss of the ear flick response when the inner ear hairs are touched. If a fish is contaminated with ciguatoxin the poison cannot be removed by washing and is not broken down by cooking.

The incidence of ciguatera in Red Sea fish is poorly documented. It is general practice throughout the region to avoid fish livers or other viscera so this helps to reduce the likelihood of such poisonings. There are however some other forms of icthyotoxin which are more certainly present in Red Sea fish and which cause problems at least as severe as those described above.

At the top of the list are the Puffer fishes such as *Arothron hispidus* and *Diodon hystrix*. In the Indo-Pacific region Puffer fish poison is well recognised and special procedures are adopted in preparation of all species from the family *Tetraodontidae* to which the Puffers belong. Dangerous species include members of the genera *Lagocephalus*, *Arothron*, *Diodon* and *Canthigaster*. Pacific islanders claim that the poison is concentrated in the ovaries of these species and that even a single egg left behind after cleaning can cause problems. In Saudi Arabia fishermen state that Puffers are safe to eat only if they are eviscerated immediately after capture. The strange aspect of Puffer fish poison is that despite their toxicity they have extremely tasty flesh which is regarded as a great delicacy in certain areas, notably Japan. For this reason there has been considerable research into the topic and very strict rules have been developed for preparation of Puffer fish. There is no evidence to suggest that Red Sea Puffers are any less toxic than those of the Indo-Pacific. They are potentially one of the most poisonous fish from the consumer's viewpoint and it is therefore worth taking a closer look at the nature of their toxicity and medical aspects of Puffer fish poisoning.

The toxin, tetrodontoxin, which has been isolated in the laboratory is an amino perhydroquinazoline with the chemical formula $C_{11}H_{17}O_8N_3$. It is slightly soluble in water and has a molecular weight of 319. The means by which the Puffer fish builds up this toxin is not clear but its development coincides with the reproductive cycle and it is concentrated in the gonads (especially the ovaries) and in the liver and intestines. The skin may also have some toxin but the body musculature itself is normally free from poison. Highest levels of toxicity occur prior to spawning at which time these fish are capable of causing the death of those who eat poorly prepared specimens. Even in Japan, where procedures are carefully laid down, there are around 30 deaths per year from Puffer fish poisoning.

The poison exerts its powerful effects on the human body

Bristly Puffer
Arothron hispidus
Photo P J Vine

by interfering with sodium transfer and thus causing blockages in the nervous system. It affects neuromuscular transmission in motor and sensory nerves and the sympathetic nervous system. In addition, it has a depressant effect on medullary centres and reduces the excitability of skeletal muscle. It also diminishes the level of intercardiac conduction and thus affects the strength of heart beat.

Symptoms of Puffer fish poison vary according to the amount of toxin ingested. Some of the first signs are weakness, pallor and dizziness accompanied by incoordination. The lips and tongue tend to tingle and this paraesthesia may gradually become more generalisd. As the toxic effects develop the patient increases salivation, sweats profusely and usually complains of chest pains or head aches. These symptoms may be accompanied by sickness (vomiting) and diarrhoea and there is often a decrease in temperature, blood pressure and pulse rate. In severe cases the skin blisters and the patient vomits blood. About 60% of those who contract Puffer fish poisoning die as a result of respiratory paralysis within 24 hours of eating the fish.

Among the various published accounts of Puffer fish poisonings there are several stories of people being assumed dead and carried to their graves while remaining perfectly aware of what was taking place! This is explained by the nature of the toxin and its effect on the neuromuscular system. At first the victim may display generalised and involuntary muscular twitching which may develop into complete paralysis. This may include a loss of speech and prevention of swallowing. The pupils initially constrict and later become fixed and dilated. In the end the victim may be unable to speak or move any part of the body and can be easily mistaken as being dead despite the fact that he is still conscious and can see what is happening!

One such case involved a Japanese gambler who ostensibly died from eating Puffer fish during the latter half of the last century. His body was placed in storage so that it could be examined by officials. Seven days later the man recovered and was able to explain that he had been aware of what was happening to him and afraid of his life that he would be buried alive! In another case the victim was placed on a cart to be taken to the next town for burial. As he was being removed from the cart he recovered and walked away. He also claimed to have realised what was happening but was unable to do anything about it.

The treatment of Puffer fish poisoning follows several lines. Among the Pacific islands, where Puffer fish toxicity is well recognised, they have a number of potions which are made from local plants and act as an emetic. This traditional treatment is supported by medical evidence and has been adopted as a standard first-aid procedure. Any method of encouraging the victim to vomit is acceptable and the literature cites inserting a finger down the throat or ingesting mustard in warm water. In cases where paralysis causes respiratory problems artificial resuscitation (see p. 111) should be carried out but care must be taken to ensure that the passage ways are clear and that the patient does not develop a bluish colour. As soon as possible, medical treatment should be sought.

Medical treatment of Puffer poisoning is discussed in appendix 1. The best way to prevent the debilitating effects of Puffer fish poisoning is, needless to say, to avoid eating Puffer fish! Inevitably, there are some who prefer dicing with death in order to taste this delicious fish. The Japanese philosopher Hosetsu Namba had this to say of his fellow-men:

"Although there are many occupations in the world, some people engage in stealing instead of entering one. Although there are many women in the world, there are some men who become adulterers instead of marrying the women. Although there are many kinds of food in the world, some people like to eat poisonous fugu (puffer fish) instead. These three groups of people make up an extremely stupid trio".

Red Sea Puffer fish are common and as a result of their rather slow cumbersome mobility they are quite easy to catch. When frightened by for example being trapped in a crevice and continually prodded by a diver, Puffer fish may rapidly swallow water and thus puff themselves up into an unwieldly but very photogenic ball. The Red Sea Puffer fish *Arothron hispidus*, which is one of the most toxic Red Sea fish to eat, has also been shown to be one of the few confirmed predators of the Crown of Thorns starfish *Acanthaster planci*. The venomous sea-star has few enemies but the two Red Sea species which have been photographed feeding on it i.e. *Arothron hispidus* and the large trigger fish *Pseudobalistes flavimarginatus* are both themselves near the top of the list of species which are, or may be, poisonous to eat. It is possible that the extreme toxicity of *Arothron hispidus* is in fact related to this food chain.

There are a number of Red Sea fish which concentrate Vitamin A in their livers. If consumed these may cause their own form of poisoning. The livers of Sharks and some of the larger groupers are especially likely to have high Vitamin A concentrations. Among the marine mammals, porpoises and dolphins are also suspect. Dugong liver on the other hand does not sem to be thus affected and is in fact very tasty and safe to eat. (Dugongs themselves are of course protected species which should not under any circumstances be hunted. On occasion however specimens are accidentally taken in fishing gear). The medical term for this form of shark or dolphin liver poisoning is "hypervitaminosis A". It is incidentally a particular problem associated with eating seal livers or those of arctic marine mammals such as Polar bears. The symptoms are similar to arsenic poisoning and they include headache, drowsiness and a general body weakness often accompanied by blurred or double vision. The patient may experience nausea, vomiting and abdominal cramps. The central nervous system is affected and a range of symptoms may include epileptic convulsions, sensory reversion, problems with speech and swallowing, and partial paralysis. Patients frequently complain of a numbness or tingling sensation in their legs. They may perspire excessively and the skin may peel or break. The heavy overdose of Vitamin A

Crown of Thorns Seastar
Acanthaster planci
Photo P J Vine

tends to stimulate bone development and there are often tender swellings on certain bones. These symptoms can last for a few days or for several weeks depending upon the extent of the Vitamin A overdose. The Vitamin is gradually excreted and serum concentration (normal level 50-100μgm %) eventually returns to non toxic concentrations. There is little first aid that can be applied since by the time the symptoms manifest themselves the vitamin A has been absorbed. The patient should be given plenty of rest and not given any foods which contain vitamin A. Medical treatment follows the same line together with the use of analgesics, tranquillisers, and anticonvulsants. Medical monitoring includes EEGs, serial serum Vitamin A estimations and bone X-rays. While this form of poisoning is a possibility it is less likely in the Red Sea than that caused by toxic shark and ray livers described below.

Sharks and rays can also cause another form of poisoning which may be confused with or combined with ciguatera and hypervitaminosis A. Shark livers are again the most dangerous organ but gonads and other viscera may also be contaminated. In this case here is a toxin present which is not destroyed by cooking but is soluble in water. For this reason preventive measures include repeated washing of shark flesh with regular changes of water together with feeding pieces of the shark or ray meat to a test animal. This form of test-feeding is widely practised among fishermen (who usually use their cat) since the symptoms of shark liver poisoning manifest themselves very soon after eating the toxic meat (unlike the case in hypervitaminosis A).

The symptoms usually start to appear less than half an hour after consumption. Initially, the victim feels sick, complains of abdominal cramps and may experience diarrhoea. Within a short space of time the afflicted person has to lie down and complains of severe headaches. Other symptoms include pains in the joints; a rapid pulse rate; tingling sensations (and possible numbness) around the mouth with a burning sensation in the throat and oesophagus; itchiness and peeling skin; together with a range of neurological and neuromuscular symptoms (e.g. weakness, incoordination, respiratory failure, double or blurred vision, lockjaw and paralysis). In severe cases victims of shark poisoning may become delirious, comatose and, in some cases, die.

Red Sea fishermen are well aware of the toxic nature of many shark livers and gonads and they avoid eating these organs. In those areas where sharks are consumed (and they are regularly sold at the local fish markets) it is important to ensure that the livers and other viscera have been properly removed and the shark meat is thoroughly washed. If, despite these precautions, shark poisoning occurs, there are a number of first-aid actions which should be followed. Firstly the victim should be encouraged to vomit since the toxic effects commence long before all the toxin has been absorbed. The standard recommended method, if the patient is conscious, is to place a finger down the throat. This may be eased by an oral dose of syrup of Ipecac (u.s.p. 8ml). In addition to treating the suffering person any other people who ate the same shark meat should be given the same first-aid. The patient should be reassured and made to rest since the symptoms may become exaggerated if the victim exercises. In acute cases mouth to mouth resuscitation may be necessary (see p. 114).

Since shark poisoning is not an infrequent occurrence in the Red Sea region, extra medical notes are provided in Appendix 2. It should be stressed that while most shark flesh is generally safe to eat and toxic effects, if any, are relatively mild, the same is not true of livers (and gonads) of Red Sea sharks which may be dangerously toxic. Islanders of Kiribati in Micronesia enjoy eating sharks which they regard as more digestible than other fish and they have a special liking for shark livers. One of their favourite foods is a shark liver sausage made by using shark intestine as the sausage casing. They are particularly wary however of the livers from tiger sharks and white-tip reef sharks and as a general rule they believe than any large specimen may have a toxic liver. Given their preference for eating livers it is hardly surprising that there have been a number of deaths caused by shark liver consumption in Kiribati. Cooper (1964) reports on three separate incidences which occurred in 1957, 1960 and 1961. The Kiribati fishermen take numerous precautions however to check on the toxicity of shark livers and they claim that there are a number of tell-tale signs including a darker than usual colour or dark spots on the poisonous livers.

Islamic teaching states that while marine creatures may be eaten by pilgrims, those animals which spend part of their

Hawksbill Turtle
Eretomycheles imbricata
Photo P J Vine

time in the sea and part on the land are not to be consumed. This ban has a great deal of biomedical logic supporting it since one of the main creatures which is thus excluded from the pilgrims' diet is turtle meat. The main Red Sea species, the hawksbill turtle (*Eretomycheles imbricata*), is capable of causing a most violent form of sea-food poisoning. While green turtles have traditionally been widely sought after for their flesh, Red Sea fishermen have long been aware that the hawksbill turtle can be deadly poisonous to eat and was valued for its shell rather than its flesh. All Red Sea turtles are

Iawksbill Turtle
:retomycheles imbricata
'hoto P J Vine

protected species by both national and international conventions and turtle consumption is therefore a rare occurrence. It is nevertheless worth commenting here upon the potentially severe consequences of eating hawksbill turtles since there is a possibility that cases will be presented in the region as a result of ill-informed people mistakenly believing that this animal is edible. The dire consequences of eating hawksbill turtles tend to be so well known among tropical marine-oriented communities that cases of poisoning are very rare indeed. One case however is reported by Cooper (1964) and concerned a group of Pacific islanders who for some reason ate this species. They all became rapidly sick and complained of stomach cramps and diarrhoea. "Their skin was hot to touch, they were unable to move their arms and legs; finally, their skin peeled off as if they had been cooked". One of the group died less than 24 hours after eating the turtle while four others died within a week of their fateful meal. In another case also reported by Cooper the symptoms were described as: "vomiting, severe stomach ache, and diarrhoea; gradual paralysis; flaking skin, leaving great sores, especially on the mouth, lips and in the armpits; intense thirst, but due to the condition of the mouth, inability to drink; finally, the victims died, described as being unable to breathe".

The contrast between the poisonous nature of hawksbill turtle flesh with the gourmet qualities of Green turtle meat is as sharp as the contrast in the feeding behaviour of the two species. While Green turtles are almost exclusively herbivorous, favouring sea-grasses, hawksbill turtles are carnivorous. The toxicity of the latter species seems to be related in some way to the food chain.

First-aid for victims of turtle poisoning follows the emetic, rest and reassurance, plus artificial resuscitation procedures described for shark liver poisoning. Medical attention and hospitalization is always required until complete recovery. Medical treatment is also similar to that given in Appendix 2.

It has recently been demonstrated that one of the most

common sea-anemones which occurs on Red Sea reefs: *Palythoa tuberculosa* secretes a toxin which is more potent than that of puffer roe. The toxicity of the anemone is related to its breeding cycle. Polyps accumulate the toxin, known as "palytoxin", in the ovaries and when eggs are released they lose their toxicity. It is been shown that some fish which feed upon the anemone may concentrate the toxin and themselves become poisonous to eat. Certain file-fish are in this category.

Having written at some length in this chapter and in the first two appendices about the potential toxic effects of eating Red Sea fish the author wishes to stress once more that such toxins are the exception rather than the rule and that the vast majority of Red Sea fish are both safe and delicious to eat. If such an account was to result in discouraging the consumption of Red Sea fish it would have failed in its main purpose which is to inform about the various toxins which can possibly occur and to report upon their treatment.

Among the thousands of invertebrate species present in the Red Sea there are a handful which can inflict painful or dangerous stings or may cause other toxic effects. The vast majority of Red Sea invertebrates are safe to touch and there is no reason why the skindiver, snorkeller or beach paddler should be afraid to enjoy the wonderful marine environment which the Red Sea offers. Fortunately the area is free from the killer box jellyfish (*Chironex fleckeri*) which so plagues the beach areas of N.E. Australia along the Great Barrier Reef coastline. There are no such deadly killers among Red Sea jellyfish and virtually all of the organisms described in this chapter are more likely to cause a painful few hours rather than any permanent disability or even death. That is not to say that killer invertebrate species do not exist at all in the Red Sea. A few deadly species are present but they are the rare exception and it is possible to avoid exposing oneself to their powerful toxins. Most serious cases of injury due to toxic marine invertebrates result from ignorance on behalf of those handling the involved species. The information provided in this chapter should help to prevent many injuries and will provide useful guide-lines concerning treatment of any injuries which do occur.

COELENTERATES

The animal phylum Coelenterata includes a large range of species which are found on coral reefs and they have a characteristic toxic apparatus which results in many of their species being able to sting divers. The phylum includes miniature fern-like hydroids; stony Millepore corals, colonial jellyfish (*Physalia*); true jellyfish, sea anemones and the main reef-building corals. The relatively simple three cell-layered structure of animals in the phylum is belied by the fact that they display a wide range of forms which are adapted to occupy many ecological niches. They are for the most part plankton feeding organisms which have specialised cells for capturing and paralysing their prey or for discouraging other

FIGURE 2
Coelenterate stinging cell (nematocyst) in undischarged and discharged modes.

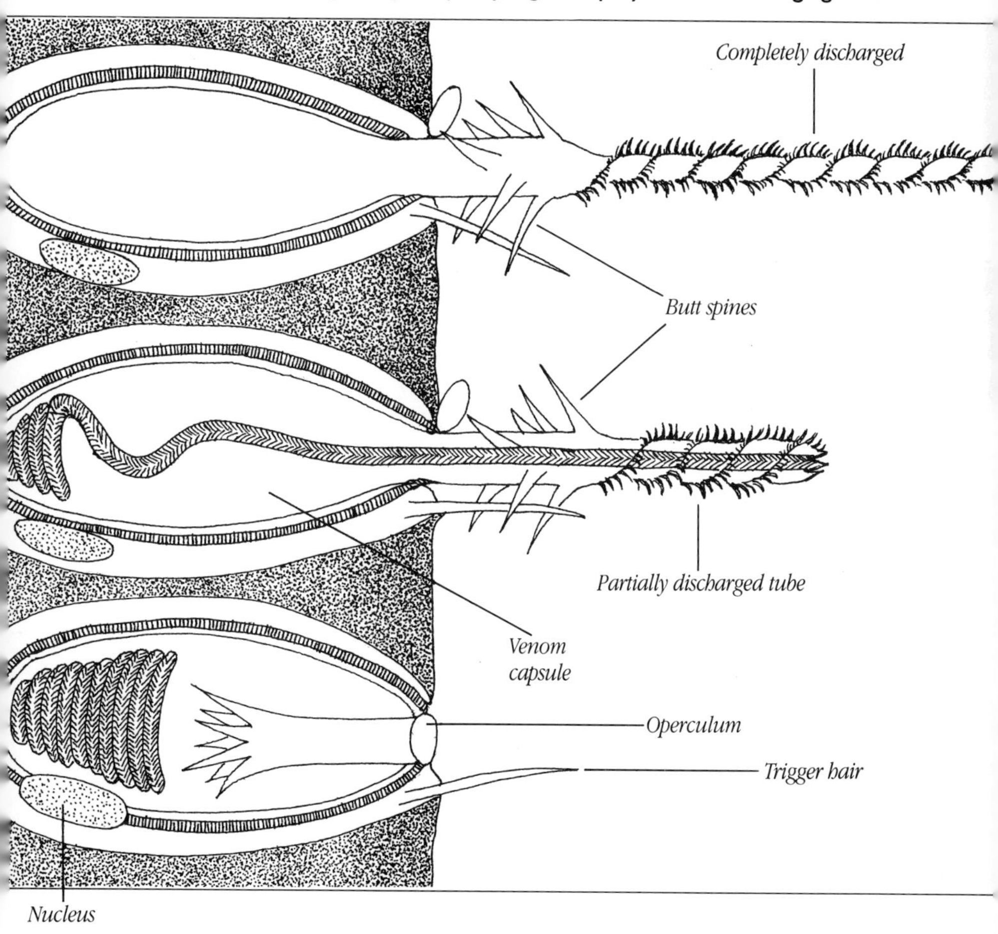

Stinging Hydroid
Photo P J Vine

animals from feeding upon them. The stinging cells found throughout the phylum are called nematocysts. There are at least three different types of these which are classified according to their various functions. So far as danger to Man is concerned it is the injector or "syringe-type" nematocysts which are of greatest interest. Such a stinging nematocyst cell is illustrated in Figure 2.

Certain Coelenterates have powerful injector nematocysts which are capable of stinging through human skin and in some cases may be able to do so through a protective covering such as a heavy gauge surgical glove. It should be noted that such nematocysts may fail to penetrate certain tough areas of skin but may do so in the thinner sections of skin such as the fore-arm or around the mouth. Injury to Man is thus caused by injection of toxin from the nematocyst capsules and also by foreign-body reaction due to penetration by the stinging threads. Reactions to Red Sea Coelenterate stings are normally quite mild but in certain extreme cases more severe effects may occur.

Coelenterate toxins are among the most potent poisons in the entire animal kingdom. The infamous box jellyfish of Australian waters can cause death of its human victims within 30 seconds. Death within two or three minutes has been regularly recorded! The chemical composition of the various toxins involved has been studied by numerous researchers but it is beyond the scope of this book to enter into detail on this subject.

Stinging Hydroids

If one examines any old mooring rope, anchor chain or similar structure which has been underwater for a while in the Red Sea it is likely that it will be covered by a fine white matting or "fuzz" of hydroids and other fouling organisms. There are several species which can inflict painful stings on unwary and unprotected divers. *Aglaophenia cupressina* and *Lytocarpus philippinus* are in this category. These worst offenders are members of the family Plumulariidae whose members may form quite large colonies. They have fan-like growth forms, frequently with a brown coloured axial skeleton. Several marine scientists have commented upon the fact that stinging hydroids vary in their degree of toxicity – being safe to handle with impunity on some occasions but able to inflict powerful stings at other times. The stinging hydroid *Lytocarpus philippinus* which occurs in the Red Sea and across the entire tropical Indo-Pacific region is greatly feared by divers in some areas (such as N.E. Australia) since they claim that repeated contact with these creatures does not lead to any immunity to their toxin in contrast to the situation with their local Sea Wasps.

A brush with a stinging hydroid such as *Lytocarpus philippinus* will usually result in a mild stinging sensation but if contact is repeated and occurs on sensitive skin areas more severe pain and other reactions may occur. The level of pain usually increases for the first ten minutes after contact. The skin which came in contact with the hydroid reddens and raised pinpoint lesions may appear. During the next twelve hours or so a generalised blotchy rash can develop and the victim may complain of itchiness. In some cases of hydroid stinging people have reported abdominal pain, cramps, diarrhoea, or fever.

Recommended first-aid procedures are as follows. Firstly, the victim should be removed from the sea, placed in a comfortable position and given reassurance. Parts of the body which came in contact with the hydroid need to be bathed with methylated spirits, or surgical spirit. Nematocysts which may be adhering to the skin may be removed by application of wet sand which is then brushed off with a towel. There is some disagreement among divers as to whether this works well since there is a danger that it will trigger more loose nematocysts to penetrate the skin. Those who support the application of sand state that it should be brushed off smoothly, rubbing in one direction only. A local anaesthetic ointment such as lignocaine 5% is best for reducing the pain.

If the situation appears to be worsening medical assistance should be enlisted. Notes on medical treatment of hydroid stings are given in appendix 3.
Although no deaths have been confirmed from hydroid stings there have been a number of cases where the cause of death was suspected to have been associated with stinging by these forms.

In general, some form of protective clothing while snorkelling or SCUBA diving will help to prevent any unwelcome discomfort from hydroid stings. If water temperatures are so warm that wet-suits are not in use an old pair of jeans and a long sleeved shirt are usually sufficient to prevent stingings caused by casual contact with hydroids.

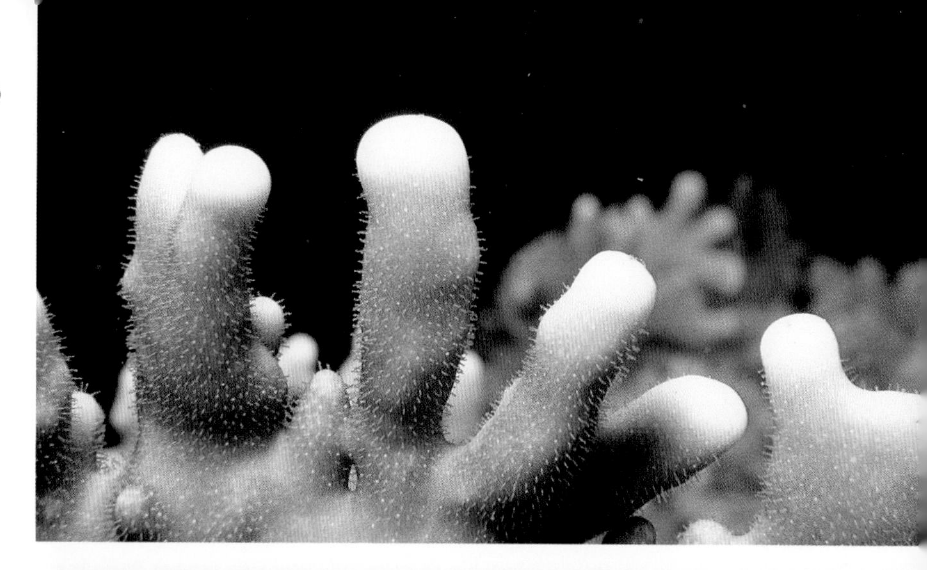

Fire Coral
Millepora
Photo F J Jackson

Yellow Stinging Coral: Millepora

In shallow water close to the reef-crest, or along the sublittoral face, Red Sea reefs have two abundant forms of yellow *Millepora* corals (also known as "fire-corals"). Intricately interwoven, bifurcated lattice-work fans are formed by *Millepora dichotoma*. These normally face into the current so that in the surge zone they lie parallel to the reef, with their broad fans facing into the wave turbulence. At greater depths, away from the continuous influence of wave-action, their fans project out from the reef and are oriented to catch plankton borne on the long-reef current.

The other form of *Millepora* is more robust and tends to form smooth colonies somewhat resembling vertically stacked plates. It is called *Millepora platyphylla* and is closely related to its more delicately shaped cousin.

Despite the coral-like appearance of both these species they do not belong to the true corals. Instead they are members of the class Hydrozoa and are thus more closely related to stinging hydroids than they are to the true corals which are members of the Class Anthozoa.

A sting from one of these *Millepora* species is normally no worse than a nettle sting but there are cases where complete ignorance of their stinging properties have led to extensive contact with the powerful stinging nematocysts and consequently quite considerable suffering on behalf of the unfortunate victims. One case which the author recalls occurred along the Sudanese coast during the mid 1970's. It involved a young Sudanese man and a Russian lady who was in the country as an interpreter for the corps of Russian military advisers who were based there at that time. Since close liaison between these two young people was banned on political and security grounds they were forced to take exceptional care about where they met.

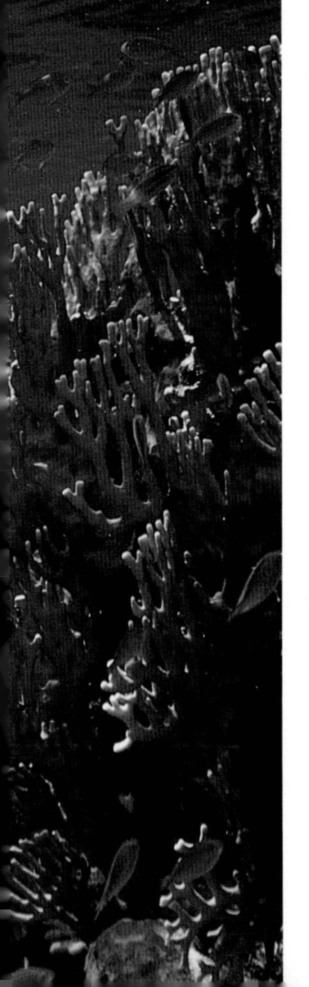

Fire-Coral
Millepora dichotoma
in the shallows
Photo A Fleetwood-Wilson

Plate Fire Coral
Millepora platyphylla
Photo P J Vine

On this particular occasion their amorous liaison was a
the reef-edge away from all public attention. Unfortunatel
they were unaware of the powerful stinging properties of
Millepora platyphylla colony which seemed to offer them
convenient seat! The result was that they were both quit
severely stung and they bore the marks of their escapade s
visibly and so painfully that it brought their friendship to
fairly abrupt halt!

During his voyage on board the Beagle, Charles Darwi
took a keen interest in coral-reefs and he became familia
with stinging Millepores. He commented as follows:

"... The stinging property seems to vary in differen
specimens when a piece was pressed or rubbed on the tende
skin of the face or arm, a prickling sensation was usuall
caused, which came on after the interval of a second, an
lasted for only a few minutes. One day, however, by merel

touching my face with one of the branches pain was instantaneously caused; it increased as usual after a few seconds, and remaining sharp for some minutes, was perceptible for half an hour afterwards. The sensation was as bad as that from a nettle, but more like that caused by the Physalia or Portuguese man-of-war. Little red spots were produced on the tender skin of the arm ...".

He thus recognised that such calcareous hydro-corals could cause varying degrees of suffering dependent upon how well their nematocysts penetrated the victim's skin. This has been remarked on by most biologists who have discussed toxicity of these organisms. First-aid and medical treatment follows the general lines given for hydroid stings (see Appendix 3). Preventive measures are also similar and include use of protective clothing together with visual recognition and avoidance of contact.

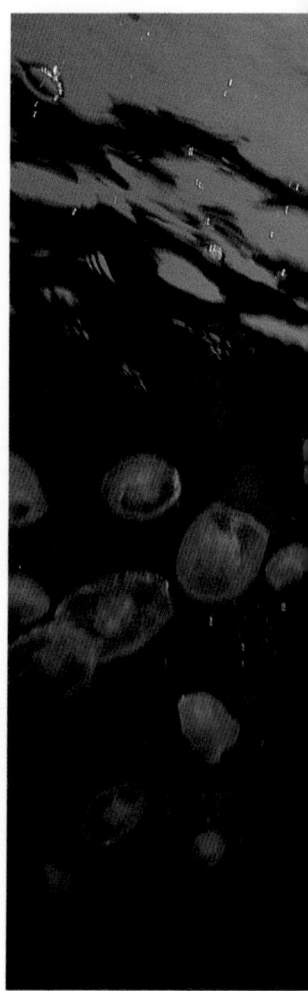

The Ubiquitous Jellyfish
Aurelia aurita
Photo F J Jackson

Jellyfish

In contrast with several other tropical marine location jellyfish do not in general pose a serious threat to Red Se swimmers and divers. One of the most frequent encountered species is *Aurelia aurita*, a ubiquitous an cosmopolitan form well known from the Atlantic and Indo Pacific as well as the Mediterranean. The main Red Se jellyfish are discussed below;

A shoal of Aurelia Jellyfish
Photo F J Jackson

Aurelia aurita

This is the jellyfish which is so well known to school children who first meet with it when it is washed up on the beach. It has a saucer shaped bell about 40cms wide with many fringe like tentacles around its periphery. While some people have been unable to elicit any sting from it, others have experienced a mild sting which may cause a slight rash which lasts for up to two hours.

The mild reaction caused by *Aurelia* is fortunate in view of its seasonal abundance in the Red Sea where, at certain times, massive numbers of the jellyfish may drift towards the reef-edge and invade shallow areas visited by swimmers. Far from being considered harmful, some have claimed that bathing in a tub-full of *Aurelia* can be beneficial as a cure for rheumatism. This practice was common in the Frisian islands during the nineteenth century.

Carybdea alata

This is one of the larger cubomedusan jellyfish and as such is related to the deadly *Chironex fleckeri* which is so feared along the tropical coastline of Australia. It does not however cause such a severe reaction as *Chironex*.

This species has been recorded from the southern Red Sea but its presence north of Kamaran island is yet to be confirmed. In view of the generally poor knowledge of Red Sea jellyfish and their distribution it would not be surprising to discover that it does occur over a wider area of the Red Sea.

Based upon observations made on it throughout its Indo-Pacific range, there are conflicting views concerning its stinging power. According to Banner (1952) this species swarms in shallow water in the Pacific coral islands of Kiribati. Such swarms are claimed by the islanders to occur one week before full-moon unless strong winds intervene to blow them onshore. Amazingly, the Kiribatans actually eat these jellyfish after scraping away the bell and tentacles and boiling the rest. If their catch exceeds their immediate requirements they simply clean them and hang them on a line to dry so that they may be stored until they are required.

The general view is that they are capable of inflicting moderate to severe stings. The weal marks are up to four in number and may be 30cms long, depending on whether or not the tentacles are contracted when contact is made with the victim. They are strong graceful swimmers which are sensitive to water turbulence and are capable of swimming at 2-3 knots for considerable distances. They tend to approach shallow beach areas when the water is especially calm.

The Upside-Down Jellyfish

Cassiopea andromeda

This species lives on the sea-bed in shallow sheltered areas, often among sea grasses. Its oral surface, surrounded by tentacles, faces upwards while the aboral surface of its bell forms a flattened concave plate providing a wide base of loose attachment to the bottom. When disturbed the bell pulsates and the jellyfish swims off for a short distance in this characteristic "upside-down" posture. *Cassiopea* can inflict quite a painful sting and the author has personal experience of this. An accidental brush with one in Port Sudan harbour caused several localized welts, a sharp pain which lasted for about 15 minutes and a feeling of general nausea. All symptoms had gone within an hour. More severe reactions have been recorded and it must be remembered that some people are more sensitive than others to such generally mild jellyfish stings.

Purple Jellyfish *Pelagia noctiluca*

This species is widely distributed through the Indo-Pacific but was originally described by Forskäl from the Red Sea. At certain times they may swarm in large numbers and make swimming an uncomfortable experience. While individual stings are not that severe, repeated contact may cause acute pain and even the collapse of swimmers.

It may be easily recognised by its hemispherical bell which ranges from rose pink to purple or yellow. The aboral surface of the bell has a warty texture and underneath this a long, thick feeding tube extends away from the bell and terminates in four long frilly feeding tentacles which surround the mouth. It gained the early attention of Red Sea explorers as a result of its conspicuous luminescence and the fact that swarms of them give the impression of glowing whiteballs moving through the water at night-time.

In Australian waters it is a source of some annoyance to surfers and has on more than one occasion been responsible for postponement of the national surfing championships!

Sanderia malayensis

This is an Indo-West Pacific species which occurs in the Red Sea and can cause quite severe stings. It has a flat disc, about 9cms wide, surrounded by long tentacles and with long frilly feeding tentacles extending from a relatively short feeding tube.

In addition to immediate pain and the development of weal marks a sting from this species may cause more generalized effects such as muscle cramps, nausea and in some instances respiratory difficulties. If a severe stinging occurs on a victim who is highly sensitive to such stings he may lose consciousness and drown. This would be most unlikely to occur however among skin and SCUBA divers who are generally aware of such a possibility and better equipped to deal with the shock of such an unpleasant experience.

Red Sea Jellyfish
Cotylorhiza erythraea
Photo J Randall

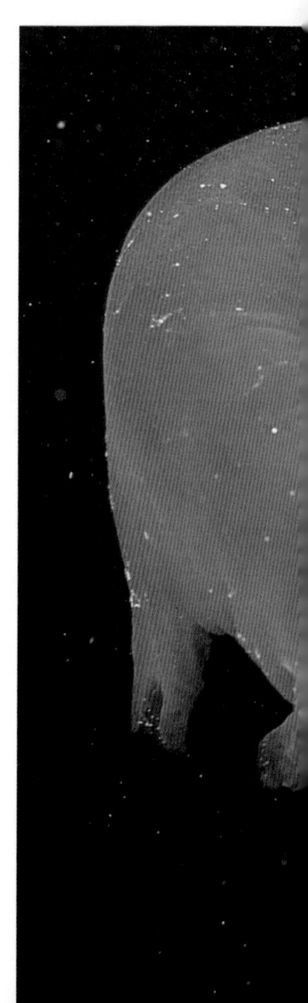

Cyanea sp

The genus *Cyanea* includes the large slimy jellyfish *Cyanea capillata* which is the largest jellyfish in the world. The species which enters the Red Sea is probably not this species but is a smaller and closely related form.

Contact with its many tentacles result in severe blistering and acute pain. If contact is prolonged the victim may experience respiratory problems, muscle cramps and other generalised symptoms. In most cases however the effects have all gone within one or two hours.

The Australian medical practitioner Dr. J.H. Barnes, who studied the effects of various tropical jellyfish described this

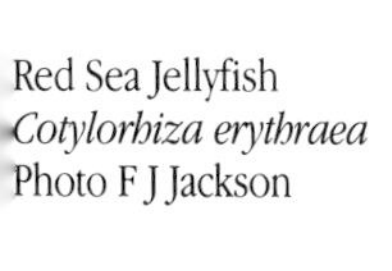

Red Sea Jellyfish
Cotylorhiza erythraea
Photo F J Jackson

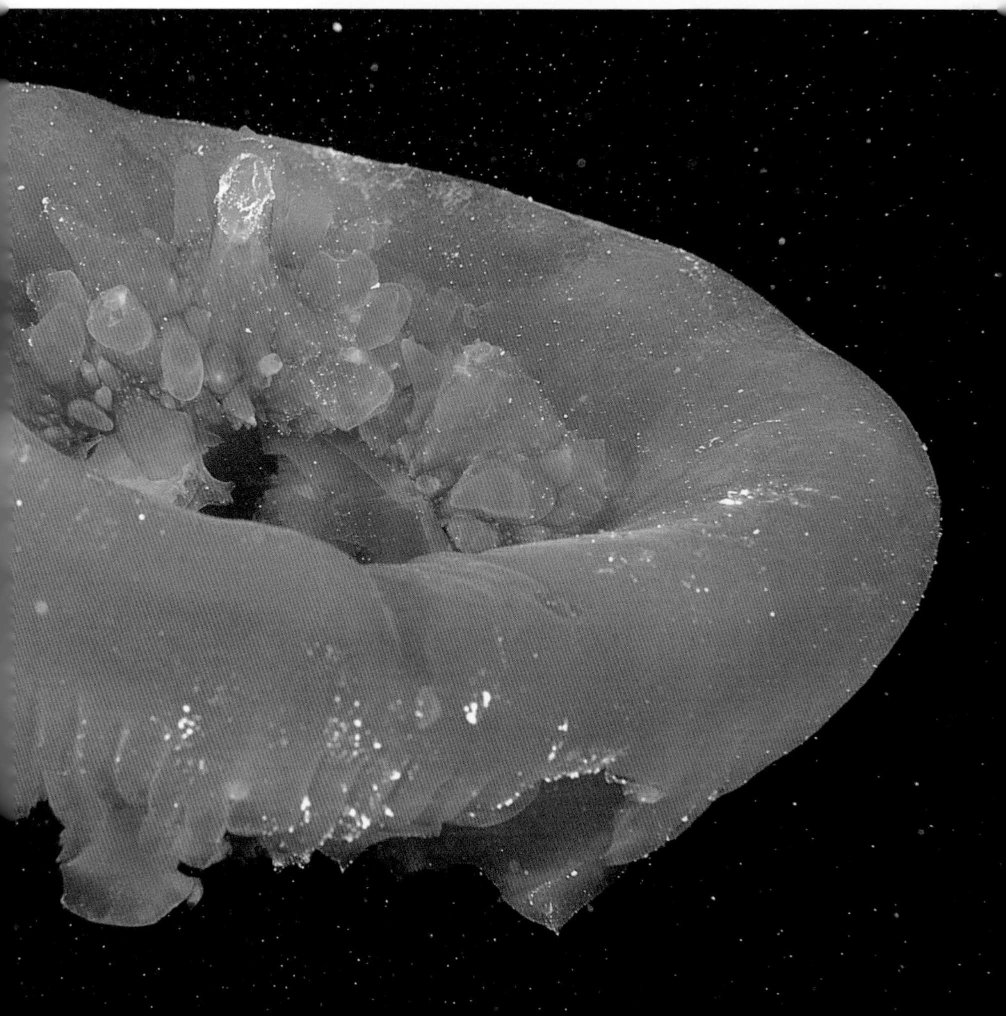

one as resembling "a mop hiding under a dinner plate". Unlike its cold water counterpart the tropical form is less of a giant with medium sized individuals measuring about 25cms across the disc. Each jellyfish possesses about 1000 well armed tentacles however with each tentacle extending to a length of about 1.5m when relaxed. Dr. Barnes recorded that in one case he registered the contraction-extension range of a single tentacle to be from 8 inches to 20 feet!

Other Red Sea jellyfish include the following Rhizostomae species: *Cephea cephea*; *Cephea octostyla*; *Cotylorhiza erythrea*; *Mastigias gracile*; *Nitrostoma caerulescens*; *Lorifera lorifera*; *Crambionella orsini* and *Rhopilema hispidum*.

Treatment of Jellyfish Stings

Reaction to jellyfish stings may range from a mild tingling sensation to a severe pain which may be associated with skin damage and more generalised symptoms such as muscle cramps or respiratory problems. There have been no confirmed deaths from Red Sea jellyfish stings.

First Aid

1 Remove victim from the sea and give reassurance.

2 Remove any jellyfish tentacles from the body. This may be done by use of sand scattered over the skin and then brushed off with a towel. Care should be taken not to increase tentacle contact with the skin since there will probably be many nematocysts which have not discharged their threads. If sand is used, it is probably preferable to use it dry rather than wet.

3 If it is available, apply liberal quantities of methylated spirits or surgical alcohol since this will help to inhibit further discharge of nematocysts which are in contact with the victim's skin. In the absence of these liquids suntan oil may be used.

4 If the pain is severe use available pain killers to provide relief.

5 If circulatory collapse follows a sting, apply standard first-aid procedures including keeping patient warm and elevating the legs.

6 If respiratory depression occurs artificial respiration may be necessary – see p. 111

7 If the sting seems to be severe and the patient's initial symptoms develop into more generalised ones medical attention should be sought. For additional medical notes see appendix 3.

Sea Anemones

Several Red Sea anemones can inflict quite painful stings to unwary divers who interfere with their tentacles. They are not generally regarded as dangerous however. Other anemones such as *Palythoa tuberculosa* may secrete powerful toxins.

MOLLUSCS

By far the most important Red Sea molluscan group from the viewpoint of their potential danger to Man are the cone-shells. There are at least thirty-eight species of this family present in the Red Sea. While all of these have venom glands, only three have been definitely implicated in serious incidents involving Man. These are the Geographic cone *Conus geographus*; the textile cone *Conus textile* and to a lesser degree the olive species *Conus lividus*. The most infamous of these is the Geographic Cone which has been responsible for a number of fatalities. One such case involved a 27 year old healthy young man who collected one of these shells while snorkelling. In order to clean off the dull periostracum he took the mollusc in his hand and proceeded to scrape at the shell with his knife. During this operation the animal stung him in the palm of his hand. The following comments are taken from an eye witness account of the events which followed (Flecker, 1936):

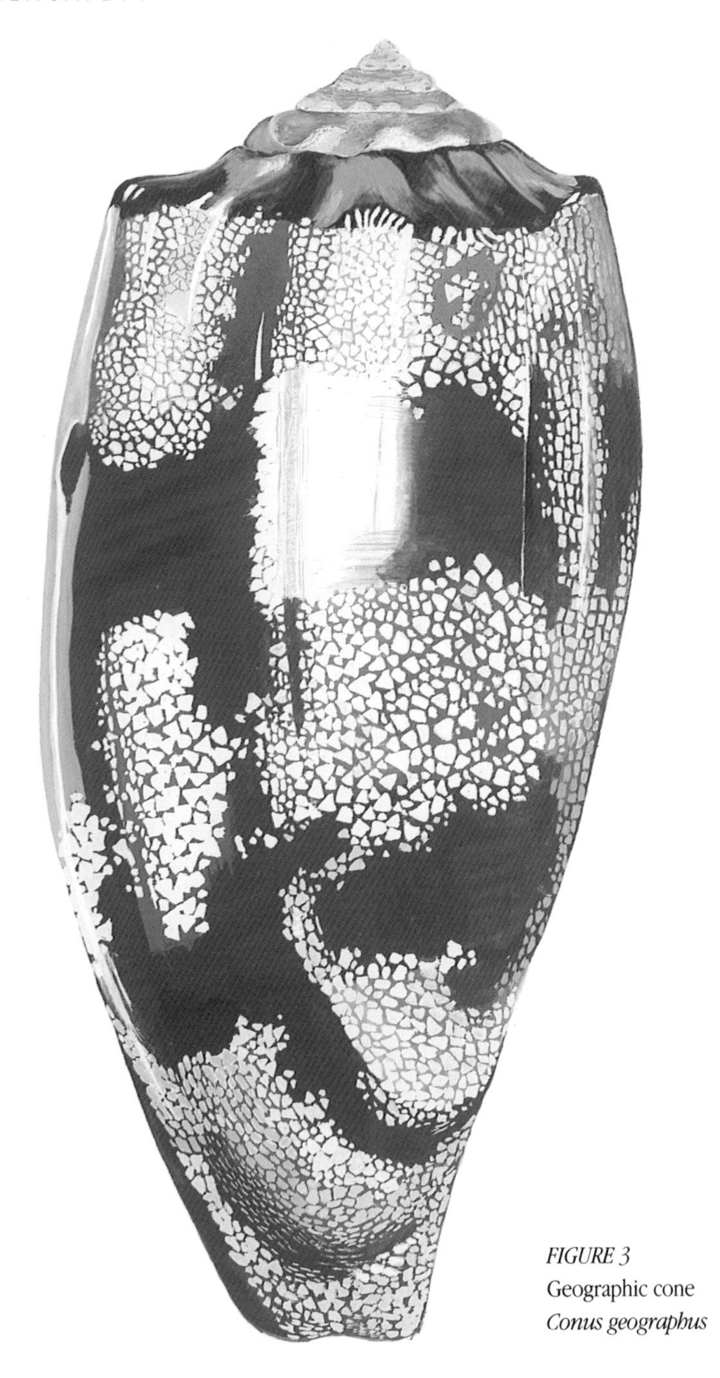

FIGURE 3
Geographic cone
Conus geographus

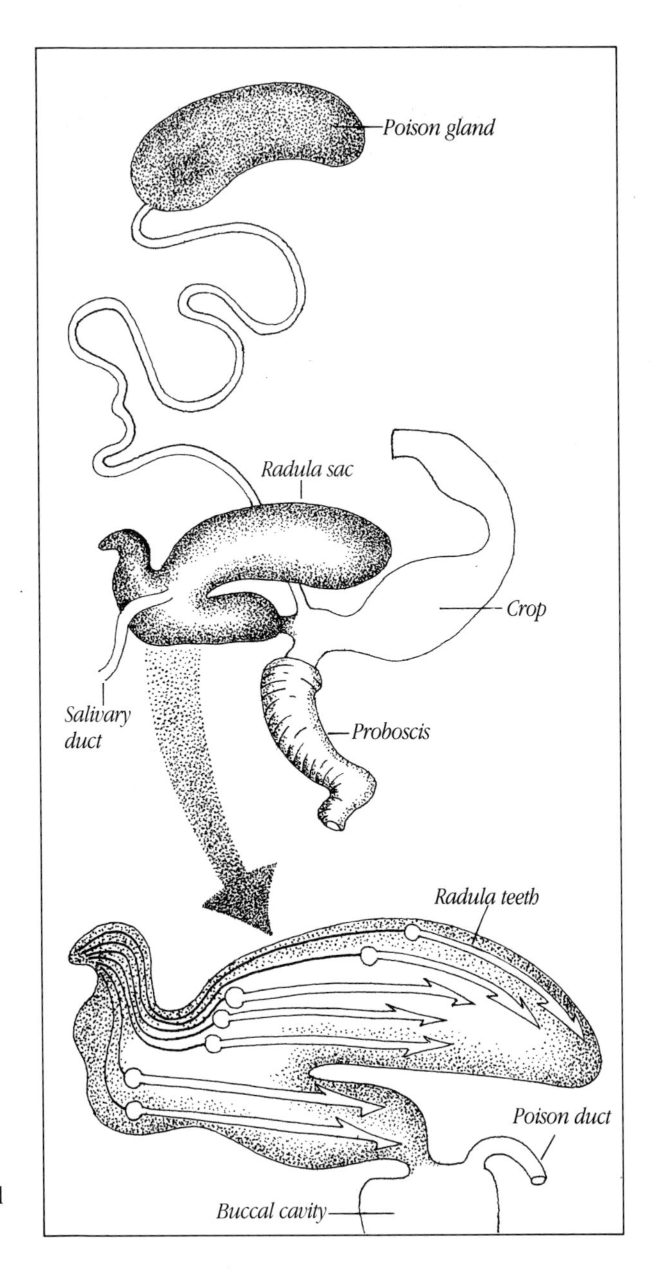

FIGURE 4
Venom apparatus of cone shell (*Conus striatus*).

Textile Cone
Conus textile
Photo J Doran

"Just a small puncture mark was visible ... Local symptoms of slight numbness started almost at once. There was no pain at any time. Ten minutes afterwards there was a feeling of stiffness about the lips. At twenty minutes the sight became blurred, with diploria; at thirty minutes the legs were paralysed; and at sixty minutes unconsciousness appeared and deepened into coma.

"No effect was noted on the skin, lymphatic, alimentary or genito-urinary systems. Just before death, the pulse became weak and rapid, with slow, shallow, respirations. Death took place five hours after the patient was stung."

In other cases of cone-shell stinging pain has been a feature of the poisoning and localised skin inflammation has also been noted. In one reported case which involved a seven year old child, both arms and legs were paralysed shortly after the sting. The child's parents managed to obtain medical assistance and artificial respiration was applied which almost certainly saved the child's life.

Cone shells live in sand or among coral rubble and they crawl along on a thick muscular foot. They tend to be nocturnal feeders and their diet ranges from marine worms and other molluscs to fish. Those which feed on fish secrete a venom which is potentially harmful to many vertebrates other than their intended prey. Man has suffered, usually as a result of shell collecting activities and careless handling of these shells.

The venom apparatus of cone shells is illustrated in Fig. 4. It consists of a venom bulb connected by a duct (which also acts as a venom gland) to the pharynx. A radula sac opening into this releases hollow barbed harpoon-like teeth from the radula and these are held in the long mobile proboscis before being pushed or fired into its prey or attacker. It has been shown quite recently that the harpoon behaves like a hypodermic needle associated with a syringe-like venom filled structure.

First Aid against cone shell stings involves removal of the harpoon if it is visible and a thorough washing of the wound. An expert on this subject, A.H. Kohn, states that the most effective early treatment probably includes "lancing of the wound and removal of as much venom as possible by suction" (Kohn, 1958). As has been stated above, artificial respiration may be necessary and normal cardiac resuscitation techniques may be indicated.

Additional medical notes on treating victims of cone-shell stings are given in Appendix 4.

Octopus

Fortunately the blue ringed octopus which inhabits Australian waters and can inflict a lethal bite is not present in the Red Sea. There are a number of Red Sea octopus species present however and all these are capable of biting a person if handled. Their salivary glands may secrete a number of toxic substances which are designed to assist in disabling their prey. There are no known reports of serious incidents involving octopus bites in the Red Sea region but it would nevertheless be prudent to take care when handling these creatures.

ECHINODERMS

Sea Urchins

Several Red Sea urchins have sharp spines which unwary swimmers or waders may accidentally touch. Their effects vary according to the species concerned and the number of spines which pierce the skin. Black spined species such as *Diadema setosum* and *Echinothrix calamaris* are the most frequently encountered urchins which cause minor injuries or on occasion quite sharp pain and temporary discomfort to those unfortunate enough to suffer multiple wounds. The major urchin species which may be regarded to a greater or lesser degree, as dangerous, are considered below.

Poisonous Sea-Urchin
Asthenosoma varium
Photo F Bavendam

Venomous Sea Urchin
Asthenosoma varium

This somewhat flattened sea-urchin actually looks dangerous to touch since its short secondary spines are covered with a thin layer of skin which forms inflated venom sacs towards their ends. When a spine penetrates a victim, it also pierces the epithelial layer encasing the spine and thus ruptures the venom sac and allows the venom to be introduced into the wound caused by the spine. This is the most poisonous of the Red Sea urchins and can inflict a very painful wound.

Long-spined sea-urchins including:
Diadema setosum, Echinothrix calamaris and E. diadema

These urchins with their long, needle-sharp spines live in sheltered reef areas. They tend to hide in holes or crevices in relatively shallow areas during the day but come out at night-time when they move over coral rubble or rock surfaces, feeding on the thin coating of blue-green algae which colonises such surfaces. *Diadema* can be locally abundant and is a frequent cause of pain to unwary divers or reef-walkers.

Sea-Urchin
Diadema setosum
Photo F J Jackson

Its spines are both needle sharp and finely barbed so that they are virtually impossible to pull out from a wound. They appear to have some mildly toxic effect which usually includes quite sharp pain and localised swelling, frequently followed by a numbness in the affected area. People who experience multiple spine penetration may also have more generalised effects of headache, and nausea but these do not usually last long.

Short-spined sea-urchin

Tripneustes gratilla

This relatively large short-spined urchin lives among sea-grass beds where it often covers itself with fragments of shells or sea grass fronds in an effort to camouflage itself. It is not generally regarded as dangerous to touch and the author has picked up many of these urchins without ever suffering any harmful experiences. This was also the experience of the Danish biologist, Mortensen, until one day he was stung on the thin skin of his arm by one of the urchins. He later realised that it was not a spine which had caused the sting but one of the tiny globiferous pedicellariae which lie between the spines and which are rarely presented with an opportunity to pierce the skin. To Mortensen's amazement the urchin's "sting" caused intense pain for several hours and the wound took over a month to heal!

The species and incident are recorded here, more as an example of how normally 'safe' animals may under special circumstances cause problems to their handlers – rather than in an effort ot have this species classified as "dangerous", which it is not.

Starfish

Acanthaster planci

There is only one species of truly venomous starfish and it is the now infamous Crown of Thorns starfish, *Acanthaster*

rown of Thorns Seastar
canthaster planci
hoto F J Jackson

planci which drew the attention of many marine biologists during the early 1970's as result of its voracious appetite for consuming live coral tissue, and its localized population explosions throughout the Indo-Pacific and Red Sea. The combination of appetite and sheer numbers was thought by many to ring the death knell for affected areas whose richly flourishing corals were such an important feature of the marine environment. Debates on cause and effect of such "plagues" continue today and are beyond the scope of this account. What concerns us here is the starfish itself and the degree of danger associated with handling it. Its entire aboral surface is covered in stout spines up to 7cms long, which are themselves enveloped in a thin layer of skin. The venom of *Acanthaster* is produced by glandular cells in the epidermis. The effects of "stings" from "Crown of Thorns" starfish vary according to the sensitivity of the person and the extent of the intoxication. The author has considerable experience of this subject having handled hundreds of *Acanthaster* while studying them in the Red Sea, and having suffered from numerous spine wounds. The effects of these wounds have generally been to cause quite sharp pain followed by localised numbness and a generalised sensation of mild nausea. After repeated wounds over a prolonged period the level of pain associated with spine penetration lessened and the mild nausea was replaced by mild euphoria!

In case I should be accused of suggesting that readers seek out *Acanthaster* as a source of a euphoric drug I should stress that while this effect has been remarked upon by other biologists and fishermen who have experienced repeated spine-wounds, the more normal response is one of severe pain and a number of associated and quite unpleasant symptoms including swelling, protracted vomiting and temporary paralysis.

Such a case was reported by Dr. J. Barnes, writing in the Australian Medical Journal in 1964. The incident involved a young Torres Straits islander who leapt out of a boat and trod directly on a Crown of Thorns Starfish. He received ten punctures in his right heel and four in his left foot. The spine penetrated about 6mm and many of them remained embedded in the wound.

The islander suffered immediate and agonising pain and although there was not much swelling the damaged areas were reddened and hot. An infection developed around the embedded spines and inguinal glands became enlarged and painful. The main clinical feature was that for the next four days or so the patient vomited – about once every four hours. Apart from this, the child was not seriously ill and after six days he was considered fully recovered.

Sea Cucumbers

Many sea-cucumbers contain a toxin known as "Holothurin". It has been analysed and identified as a saponin and is generally concentrated in the body wall; viscera and Cuverian tubules. The toxin has a direct effect on muscle contraction and has a nerve blocking action similar to that exhibited by cocaine, procaine and physostigmine on laboratory animals. Its effects on humans have yet to be clarified but it is reported that liquid ejected from the viseral cavity of some species may cause dermatitis or blindness in man. Ingestion of sea-cucumber poison is also claimed to be potentially lethal but reported cases of sea-cucumber poisoning are rare. Acute conjunctivitis has been reported in persons swimming in water where tissue extracts of sea-cucumbers are known to be present. Such incidents are usually linked with the use of holothurians to poison fish on coral-reefs. Fishermen have learnt that mashed-up sea-cucumbers spread over the water in pools will render fish less active and easier to catch.

The role of sea-cucumber toxins is probably to protect them from predators. The common black sea-cucumber, *Holothuria atra*, releases copious quantities of a red coloured toxin when handled. This will kill or render inactive many small invertebrates which are in the vicinity and is an effective deterrent against predation.

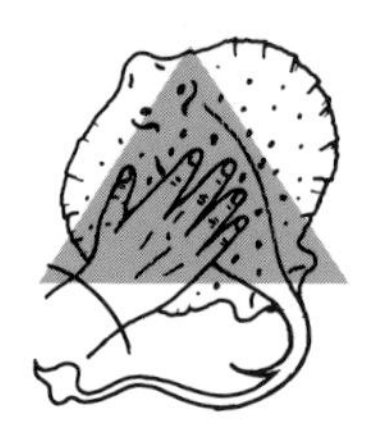

First of all let us deal with Sea-snakes. Despite the fact that there are some snake-like eels which have given rise on occasion to reported sea-snake sightings in the Red Sea, it is generally held by marine scientists that sea-snakes do not occur in the Red Sea. After thousands of hours spent in and around the Red Sea the author is satisfied that this is the case and that it is extremely unlikely that any diver or reef-walker will be faced with the danger of venomous sea-snake bites in the Red Sea. The question of *why* sea-snakes are *not* present is an interesting one since they occur in the Indian Ocean and are positively common in the Arabian Gulf. I put this question to Dr. George Pickwell of the U.S. Navy who has studied Indo-Pacific sea-snakes and has prepared a Naval Manual on various dangerous animals. He commented as follows:

"In the matter of the lack of sea snakes in the Red Sea – my own best guess as to why this should be so is that they find insufficient prey items for their needs. As you say, there is likely too little continental shelf for suitable prey habitat. The great majority of sea-snakes dive for their food and the greatest verified depth to which they go in search of food is in the order of 50 metres, although most, I suspect, do not dive nearly that deep. In other words, a shallow water habitat for the eels and other prey items they consume is a must. Still, not even *Pelamis* is taken in the Red Sea and it is a surface feeder. So perhaps the true limiting factor is a compound of several different features of sea-snake biology. In any case, it is an intriguing facet of general sea-snake biology".

It may be intriguing to the scientists but it is more of a relief to the general public who for some as yet undetermined reason are not faced with the danger of encountering sea-snakes in the Red Sea.

So, with that bogey out of the way, what is left to fear among the numerous Red Sea fish which one may meet while wading, swimming, diving or perhaps fishing in the area? There are a number of quite dangerous species of which one should be aware and which should, if possible, be avoided.

eef Stingray
aeniura lymma
hoto F J Jackson

STINGRAYS

Reef Sting Ray

Taeniura lymma

The most frequently encountered ray in shallow waters among Red Sea reefs is the blue-spotted Reef Stingray – *Taeniura lymma*. During the daytime it can often be seen resting on the sandy bottom of sheltered lagoons under the shade of coral ledges while in the early morning or evening it ranges out across the lagoon floor in search of its bottom-living food which consists mainly of burrowing molluscs,

Reef Stingray
Taeniura lymma
Photo F J Jackson

crustaceans and other sedentary creatures. Its venom apparatus consists of one or two venomous spines extending from the mid-posterior part of the disc. The spines (figure 5) have a series of recurved barbs rendering them very difficult to withdraw from a wound. The venom glands lie in a pair of longitudinal grooves along the upper edges of the spine. If trodden on or grasped, this ray is capable of thrusting its tail upward and forward and driving its spine into the body of the victim. The thin layer of skin covering the spine is thus ruptured and venom then flows along the longitudinally grooved spine into the wound. Spine penetration by this and

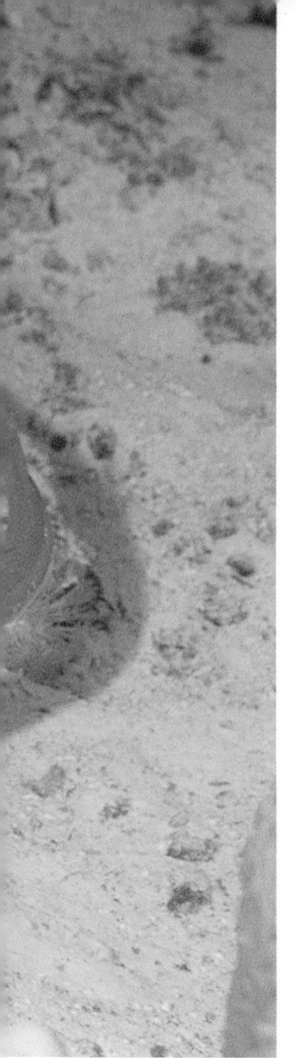

other rays is reported to be extremely painful and can result in quite severe symptoms.

The venom is a high molecular weight protein which is regarded as lethal (LD_{50} is around 300mg/kg body weight). Under normal circumstances, however, much less venom than this is injected into the wound and lower concentrations cause effects ranging from pain, nausea and muscular cramps to cardiac irregularities, breathing difficulties and fever. Snorkel and SCUBA divers are fairly unlikely to be stung by this species however since it never adopts an aggressive attitude to divers unless it is actually caught or trapped in some way. Accidents have occurred however as a result of inadvertent contact with rays which have been buried in the sand or caught in fishing nets.

First Aid treatment involves placing the patient in a resting position with the affected portion of the body higher than the rest. The wound should be thoroughly washed, with sea-water if necessary, to remove all surface venom. The spine should be extracted as carefully as possible since its recurved serrations will probably cause further laceration. If bleeding is not already occurring, a small cut should be made to encourage bleeding. The venom is denatured by heat and therefore the wounded area and adjacent parts of skin should be immersed in water at 50°C until the pain ceases (usually within 2 hours). One should be aware that stings can result in severe symptoms and that medical attention should always be sought following the initial first-aid.

Additional medical notes on treatment of stingray wounds are given in Appendix 5.

Stingrays often lie buried in the sand and stings may result from them being trodden on. Fig. 5 shows how such a sting occurs. If one is forced to wade through shallow areas inhabited by stingrays it is a good idea to prod the sand in front with a stick and to shuffle one's feet rather than take long striding paces through the water. In addition, extreme care is necessary when handling rays which have been caught in fishing nets or on lines.

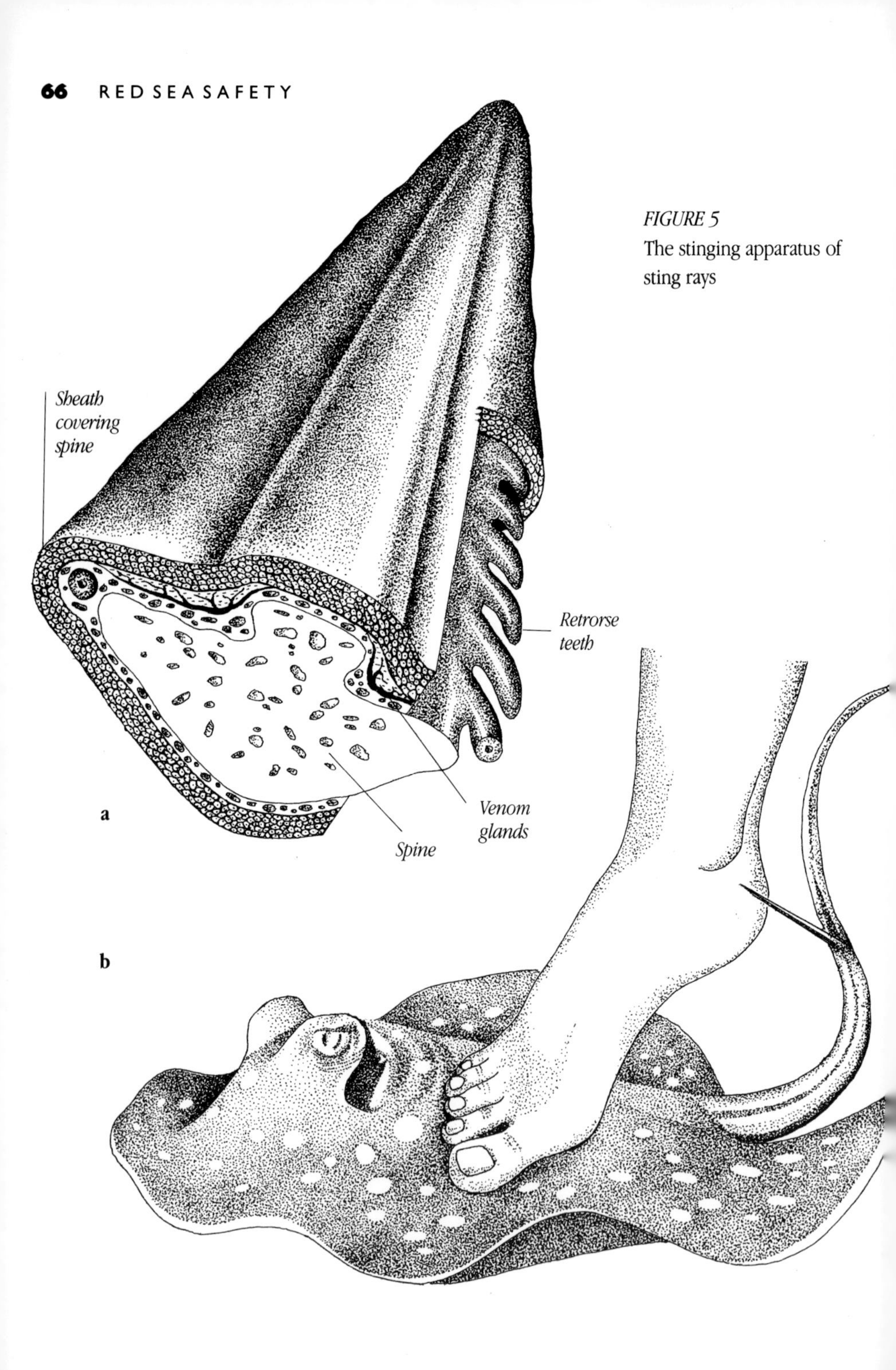

FIGURE 5
The stinging apparatus of sting rays

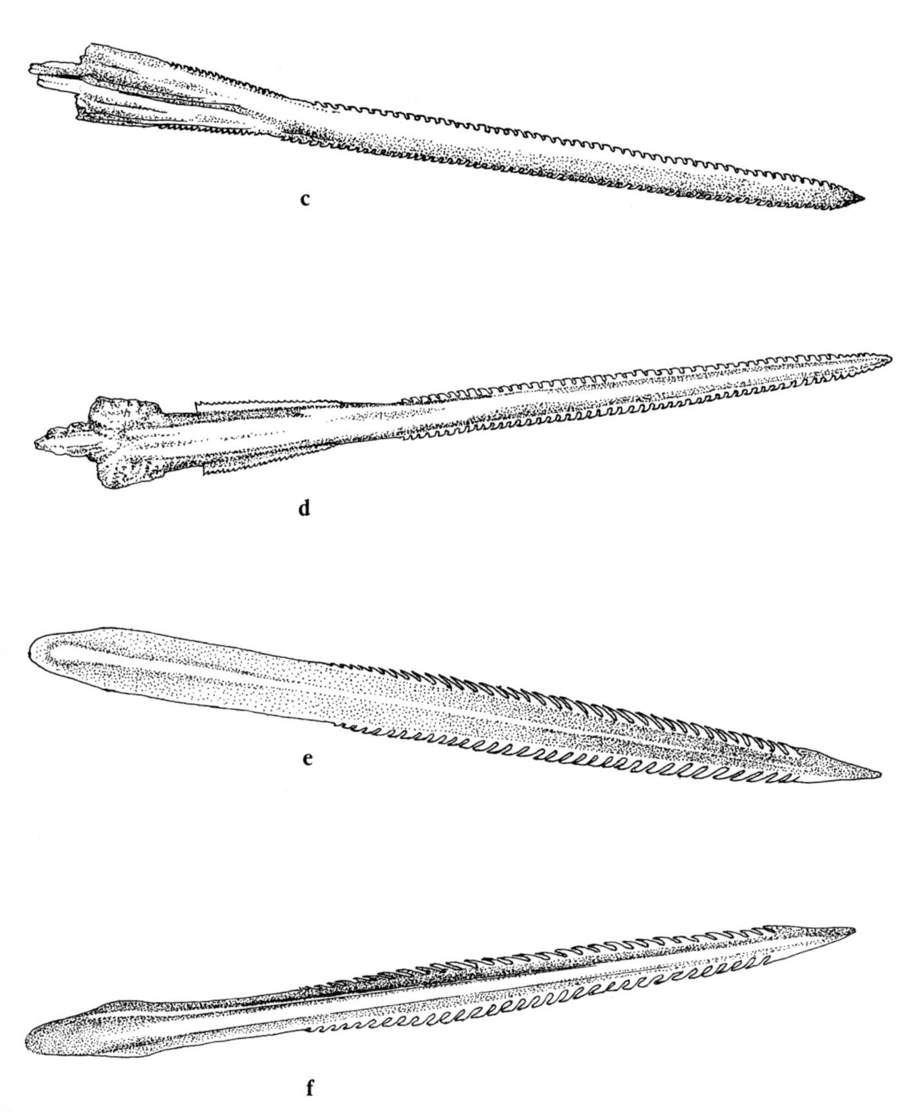

Gross anatomy of typical stingray sting

Method of stinging – *Taeniura lymma*.

Spine of *Aetobatus narinari* – ventral view.

d Spine of *Aetobatus ṇarinari* – dorsal view.

e Spine of *Taeniura lymma* – dorsal view.

f Spine of *Taeniura lymma* – ventral view.

Himantura uarnak

This spotted stingray frequents shallow areas and can be found for example along the shores of Suakin harbour where it is quite common. Red Sea fishermen are aware of the danger of being stung by this ray and they take care to avoid stepping on it when cast-net fishing for bait.

(For notes on medical treatment see Appendix 5).

Aetobatus narinari

This large ray occurs in deep water, along the reef-face, as well as in the shallows where it sometimes feeds in muddy lagoons. It has a long venom spine (with backwardly directed serrations) near the base of its tail and this is covered by a thin layer of skin containing venom secreting cells. Despite the spine position at the tail base, the ray is able to curl its tail forward and strike with its spine. There have been several incidents of this happening to unsuspecting bathers in Red Sea coastal waters.

ELECTRIC RAY

Torpedo fuscomaculata

On two occasions, while taking underwater photographs, I have sat or knelt on large electric rays. The effect was instantaneous, unexpected and highly stimulating! A powerful electric current sent me reeling backwards and left me for a moment in a state of dazed shock. On both occasions the encounter took place around 20 metres deep on a sandy sea-bed, close to coral outcrops. The smooth sand covering the back of the ray had looked like a good place to rest on the bottom while focussing the camera. Simultaneously with the electric shock the sea-bed erupted in a flurry of sand and beating of "wings" as the ray protested at such a rude intrusion on its siesta.

Electric rays, such as the Red Sea species, *Torpedo fuscomaculata*, are capable of delivering a short burst of

Electric Ray
Torpedo fuscomaculata
Photo F J Jackson

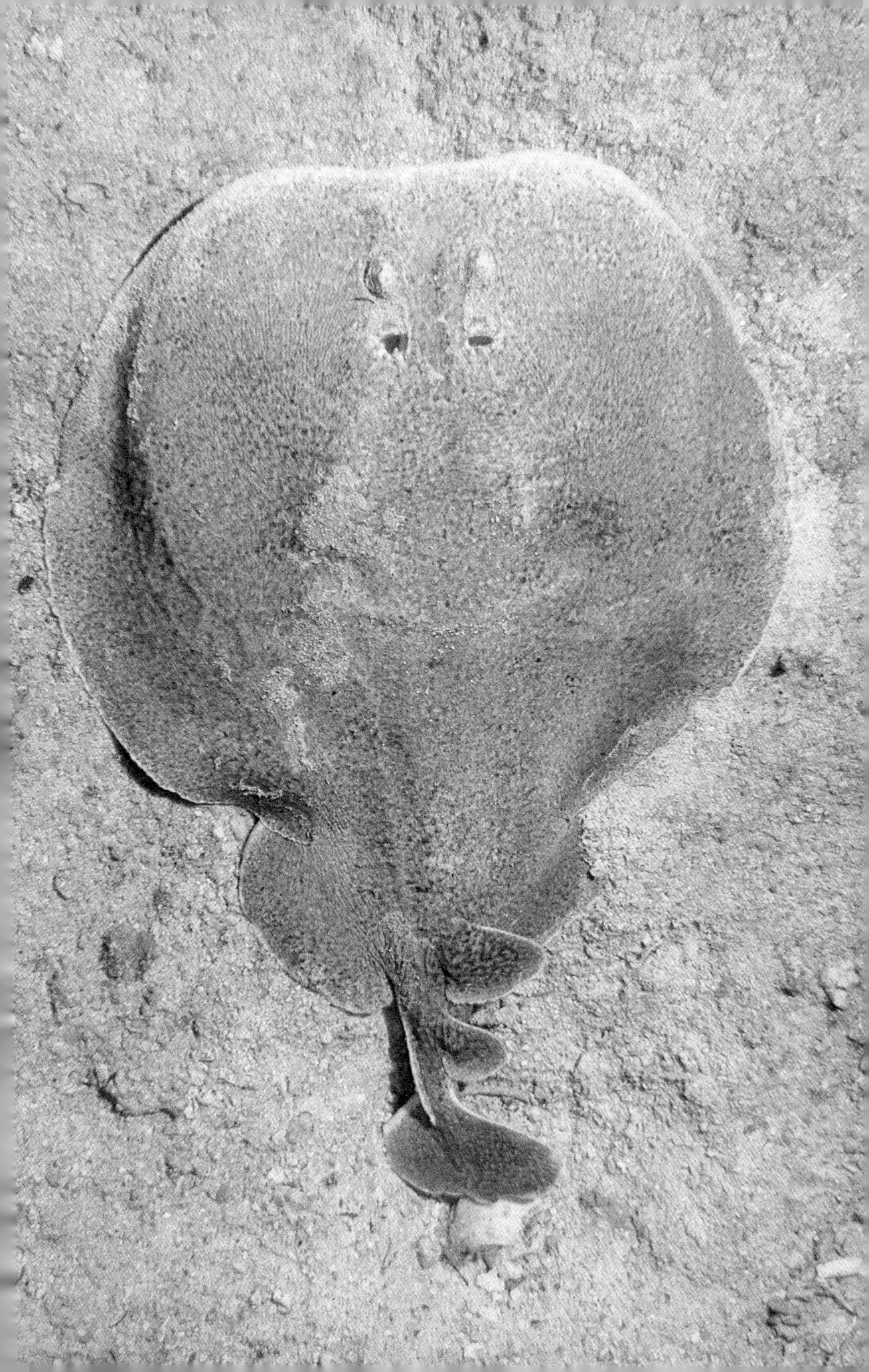

Electric Ray
Torpedo fuscomaculata
Photo F J Jackson

electricity as strong as 200 volts and 2000 watts. They use their electricity as a defence system and also as a means of stunning their prey. Their electric organs are situated at the front of the disc – one on each side, between the forward edge of the pectoral fin and the head. The electric charge passes between the negatively charged ventral side and the electrically positive dorsal side of the ray. The output of electricity is a reflex action stimulated by touching the ray and it is not necessary to touch it on both surfaces. The initial discharge is the strongest and subsequent shocks are less powerful. After repeated electrical discharge there is a period when the ray lacks much electrical energy and is, so to speak, "recharging its batteries". They tend to be feeble swimmers and are thus relatively easy prey to sharks which feed on them despite their impressive electrical defence mechanism. The electric ray itself feeds on benthic organisms including crustaceans, molluscs, worms, other invertebrates and some fish.

The main danger from such Electric Rays is that of being virtually knocked unconscious underwater. It is unlikely to happen however and in the case of my own contact with large specimens, the shock was more akin to being "booted in the backside" – suddenly and unexpectedly – than to a more serious or dangerous jolt.

The strange power of *Torpedo* has been known since ancient times. It was Aristotle who wrote that it ejected "a sort of poison or elixir being neither" – a confusion caused by the fact that electricity was at that time unknown.

TURKEY FISH OR LION FISH

Pterois volitans and Pterois radiata

On first encountering a lion fish drifting above the reef with all its beautiful fins extended forward, and in the characteristic head down posture, one's immediate response might be to swim over and touch this graceful creature. It would, to say the least, be unwise to do so! Those feather-like fine rays have the capability to inject a powerful toxin into anyone who meddles too closely. Its bright colours and elaborately developed fins serve the dual functions of, on the one hand, providing a disruptive camouflage which confuses their prey (since they look more like a drifting fragment of weed than a voracious and efficient predator); and on the other hand broadcasting a message of "hands-off" to any potential predator which might otherwise feel that their strange form is accompanied by a less than efficient escape response. Although they may not be the fastest long distance swimmers on the reef they can put up an incredibly rapid lunge towards their prey aided by opening the mouth wide and literally sucking in water. Their unsuspecting prey usually disappear in one gulp – almost too fast to notice by eye.

Turkeyfish stings generally occur to people who are unaware of the danger posed by this fish. Also, occasionally, fishermen who capture them in their nets are stung while trying to remove them. There is at least one recorded fatality

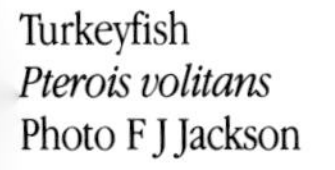

Turkeyfish
Pterois volitans
Photo F J Jackson

Clearfin Turkeyfish
Pterois radiata
Photo F J Jackson

from *Pterois* in the Red Sea; that of a chld who was stung while swimming in Sharm Obhur north of Jeddah. Death occurred three days after the event.

The spines are covered by a thin layer of skin which ruptures when they penetrate a victim and the venom then flows along the spine(s) and into the wound(s). Usually a single spine penetrates and causes a puncture wound of 2-5mm in diameter. This soon develops a red or blue "halo" while surrounding tissue becomes pallid and swollen. The immediate effect of such an incident is that of severe pain which increases over the next few minutes and may be so excruciating as to cause considerable distress. The pain may diminish after several hours or it may continue for one or two days.

First Aid procedure requires that the patient is laid down with the wounded part of the body elevated above the rest. The wound should then be washed – using copious quantities of sea-water. Any foreign bodies should be removed from the wound and if it is not bleeding a small incision across the wound and parallel to the long axis of the limb may help to clear the wound and relieve some of the pain.

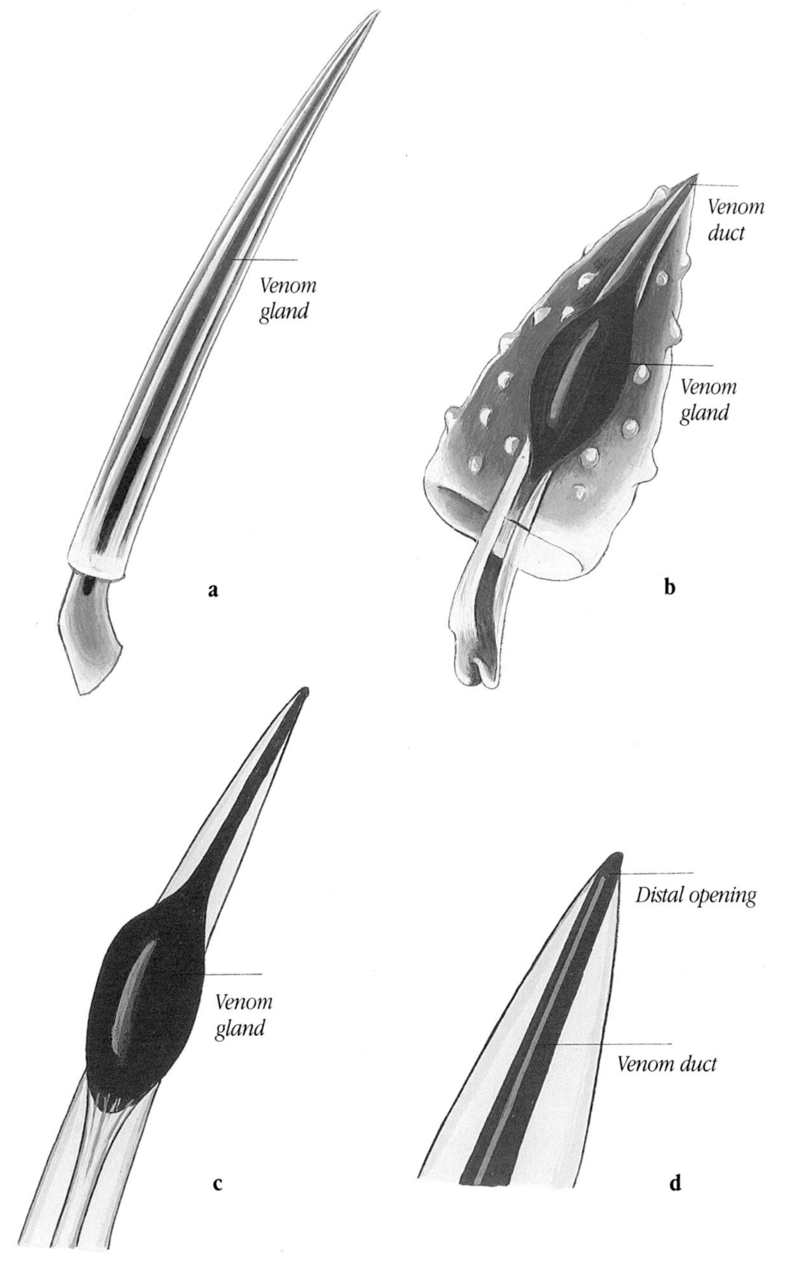

FIGURE 7
Morphology of stings of
a Turkeyfish (*Pterois*)
b Stonefish (*Synanceia verrucosa*) showing covering of skin.
c and **d** Stonefish dorsal spine dissected out from fish.

Illustration Jane Stark

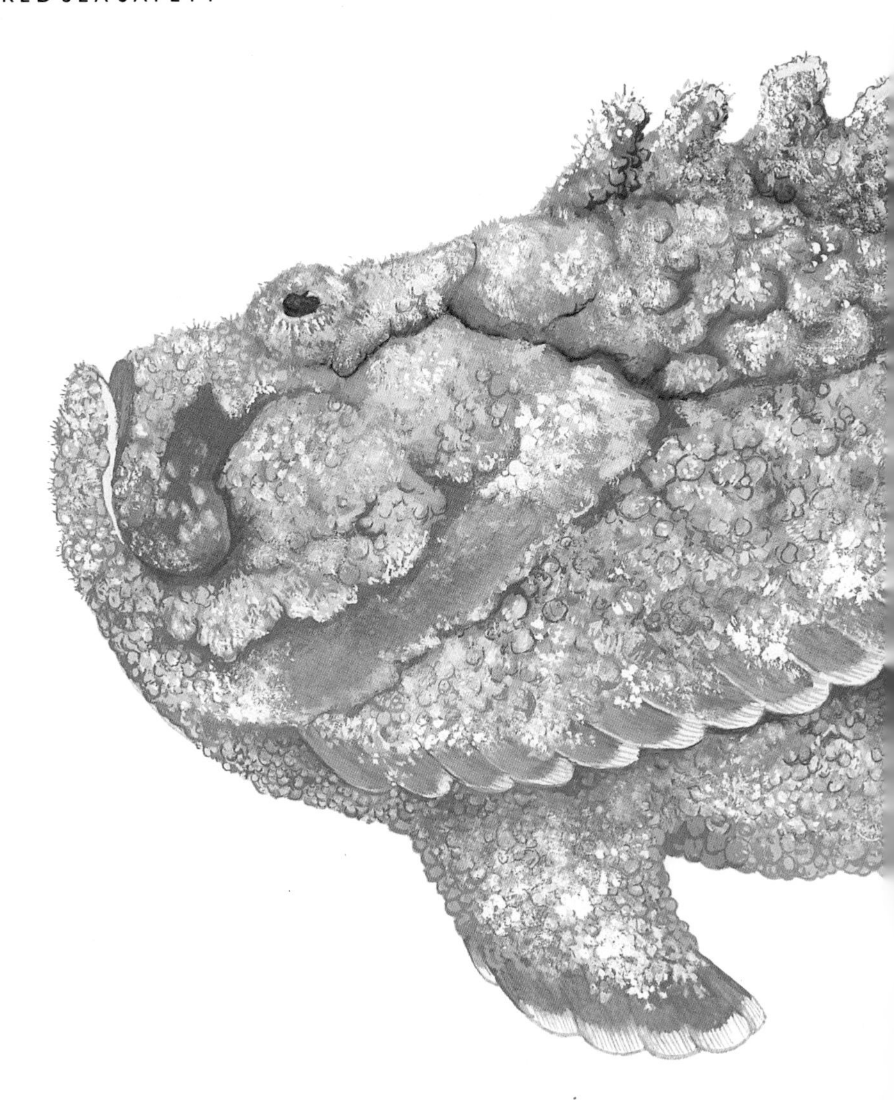

The venom is heat sensitive and the affected part of the body should therefore be placed in, or bathed with, hot water at 50°C. This will relieve pain and should be continued until there is a considerable improvement.

Only when this initial assistance is being, or has been, given should time be taken to summon medical assistance. Further medical notes are given in Appendix 6.

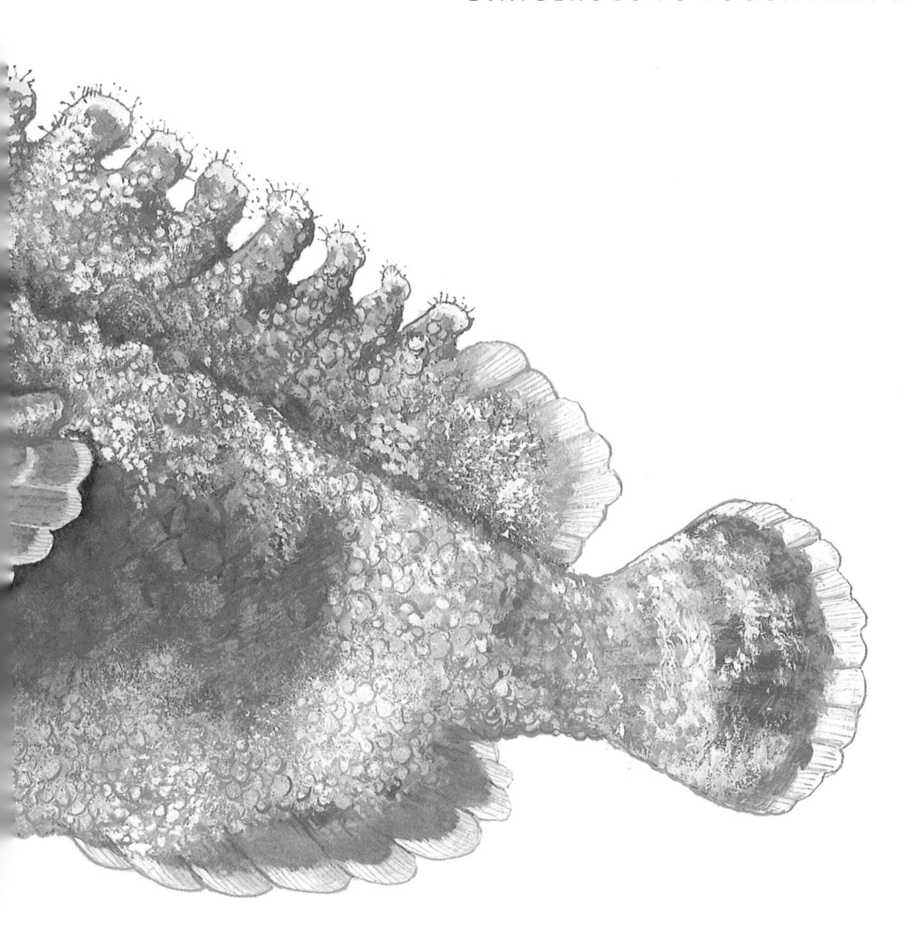

FIGURE 6
Stonefish: *Synanceia verrucosa.*
Illustration Jane Stark

STONE FISH

Synanceia verrucosa

This is probably the most dangerous fish in the Red Sea since it is extremely well camouflaged – to the point of being virtually invisible; among coral rubble or stones (often close to artificial quays where people frequently wade); has needle sharp dorsal spines capable of penetrating a rubber soled shoe and it can inject a potentially lethal venom.

Stonefish have the ability to alter their colour pattern so that they merge almost perfectly with whatever their surroundings may be. When they are trodden-on the dorsal spines penetrate the foot and the sudden weight on the fish compresses the poison glands which discharge their venom through the longitudinally grooved spines and into the wound (Fig. 7).

The author has seen two people stand on stonefish. In one case the spines penetrated through strong polythene sandals and when the victim raised his foot out of the water the stonefish was still suspended from the sandal. In the second case, which occurred at Sharm Obhour – "the creek" – north of Jeddah, a fisherman jumped into the water from a stone quay and landed on top of a stonefish! Both the observed victims survived despite poor medical treatment in the first case and delayed treatment in the second. Nevertheless, the incidents have left a deep impression of the sheer agony inflicted by stonefish stings.

Stonefish venom acts as a myotoxin on various muscles including skeletal, involuntary and cardiac muscle. The results can entail muscular paralysis, respiratory depression, peripheral vasodilation, shock and cardiac arrest. Each stonefish dorsal spine has 5-10mg of venom associated with it. The venom (pH 6.8) is an unstable high molecular weight protein (MW 150.000) which is denatured by heat (2 minutes at 50°C) and by both alkalis and acids (i.e. pH greater than 9 or less than 4). Potassium permanganate is also said to destroy it.

LEFT AND LOWER RIGHT
False Stonefish/Scorpionfish
Scorpaenopsis gibbosa
Photos F J Jackson

ABOVE
Stonefish
Synanceia verrucosa
Photo F Bavendam

A stonefish antivenom has been produced by Commonwealth Serum Laboratories, Melbourne, Australia. (See Appendix 7 for address and ordering details). It consists of refined and concentrated hyperimmune horse serum such that 1ml of the antivenom neutralises 10mg of stonefish venom. The initial recommended dose is 2ml of antivenom administered intramuscularly. This may be followed by a further dose of 2ml if symptoms persist. The Commonwealth Serum Laboratories recommend that a supplementary pain relieving treatment such as lignocaine is administered and that emetine hydrochloride (65mg/ml) or a few minimes of 1-1000 solution of potassium permanganate are injected into the sting site. These treatments will destroy any venom which remains at the site of spine penetration.

Before one is able to carry a stonefish victim to a doctor who stocks antivenom there is a great deal that can be done to alleviate the pain and to increase the chances of survival. Hot water should be applied to the wounded region as soon as possible. The wound should be cleaned and any foreign bodies removed. The patient should be lain down with the affected limb raised above the rest of the body. Although some authors have recommended use of a ligature this can cause more problems for the patient and one should only be applied by a qualified paramedic in cases where the victim is treated within a minute or two of the incident or when death seems imminent. An incorrectly applied ligature may be more dangerous than the stonefish sting! If potassium permanganate is available a weak (5%) solution should be used to irrigate the wound. In different regions of the world locally available natural cures have been developed. The Australian aborigines for example use mangrove sap.

By far the most reliable first-aid measure in cases of stonefish stings is the hot water bath mentioned above. This should be maintained for about an hour, or even as long as an hour and a half. The water should be at around 50°C and surrounding skin areas should also be immersed to prevent scalding.

Finally, if the patient looses consciousness external cardiac massage and respiratory resuscitation should be applied (see p.113). This should *not* be stopped until a medical practitioner is present to advise further, or if the patient recovers. Medical notes on treating victims of stonefish stings are given in Appendix 7.

Scorpion Fish
Scorpaenopsis gibbosa
Photo F J Jackson

Published with kind permission of Commonwealth Serum Laboratories

Stonefish antivenom

This antivenom is prepared by hyperimmunizing horses with the venom of the stonefish *(Synanceja trachynis)*. The hyper-immune serum is refined, concentrated and standardized, and Phenol 0.25% w/v is added as a preservative.

One unit of Stonefish Antivenom neutralizes 0.01 mg of stonefish venom (1000 units will neutralize 10 mg of venom). On the average, 5-10 mg of venom is contained in each of the 13 dorsal spines of the stonefish.

Indications for use

Stonefish Antivenom neutralizes all the toxic effects of stonefish venom and is indicated in all cases of stonefish sting.

The dominant symptom of stonefish sting is agonizing and persisting pain. Oedema, which usually develops rapidly after a sting, may become extreme. Abscess formation, necrosis and gangrene have occurred in untreated cases. In addition to the local effects, muscle weakness and paralysis may develop in the affected limb and varying degrees of shock may occur. The systemic effects are due to the presence of potent myotoxins, which act directly on all types of muscle – skeletal, involuntary and cardiac.

Dosage and administration

The initial dose will depend on the number of punctures visible, eg.

1 or 2 punctures – 2000 units (contents of 1 ampoule)
3 or 4 punctures – 4000 units (contents of 2 ampoules)
5 or 6 punctures – 6000 units (contents of 3 ampoules)
and so on.

The antivenom should be given by intramuscular injection, but in severe cases should be given by intravenous infusion. If symptoms develop or persist and the identity of the stonefish is assured, the initial dose should be repeated.

Precautions

The usual precautions in the administration of heterologous antiserum should be observed.

Supportive therapy

Local infiltration of the wound with emetine hydrochloride (65 mg/ml), lignocaine or potassium permanganate (5%) may help to alleviate pain. Immersion of the limb in hot water has sometimes brought relief when other measures have failed, but care must be taken to ensure that the patient is not scalded.

Mode of issue

Stonefish Antivenom is issued in containers of 2000 units (approximately 2 ml).

Storage

Antivenoms should be stored, protected from light, at 2° to 8°C. They must not be frozen.

Record of cases

Since relatively few reports have been received about stonefish sting and its treatment, it would be appreciated if practitioners complete and return the accompanying questionnaire.

Reference

Wiener, S., "Stonefish Sting and its Treatment." Med. J. Aust., 2: 218-222 (1958).

anuary, 1980
60.0000

Shortfin Lionfish
Dendrochirus brachypterus
Photo F Bavendam

Other Scorpionfish

There are about 35 Red Sea scorpionfish but the most frequently encountered species are Lionfish or Turkeyfish (Pterois), stonefish and a few others including the False Stonefish (*Scorpaenopsis gibbus*), the Bearded Scorpionfish, (*Scorpaenopsis barbatus*), the Devil Scorpion (*S. diabolus*) and the Shortfin Lionfish (*Dendrochirus brachypterus*). In all these fish the -dorsal, anal and pelvic fins are venomous. Stings may vary in severity from something resembling a bee sting to an agonisingly painful experience which can, under certain circumstances, culminate in death of the victim.

First-aid treatment is similar to that described for lionfish and stonefish – i.e. basically cleaning the wound, application of hot water, elevation of the affected limb, local and general anaesthetics, antibiotics, and tetanus prophylaxis. There are scorpionfish antivenoms produced and if these are available they may also be used.

Orangespine Unicornfish
Naso lituratus
Photo F J Jackson

SURGEON FISH
Acanthuridae

Most texts on dangerous marine animals include comments on Surgeon fishes. This author is of the opinion that the real danger to swimmers from Red Sea Surgeonfish is so remote as to be virtually non-existent. Nevertheless they have a reputation as potentially harmful creatures and we shall briefly examine their credentials in this sphere!

There are at least eleven species of the family Acanthuridae in the Red Sea. They are listed below:

Sailfish Surgeonfish *Zebrasoma veliferum*
Yellowtail Surgeonfish *Zebrasoma xanthurum*
Black Surgeonfish *Acanthurus nigricans*
Brown Surgeonfish *Acanthurus nigrofuscus*
Sohal *Acanthurus sohal*
Bleeker's Surgeonfish *Acanthurus bleekeri*

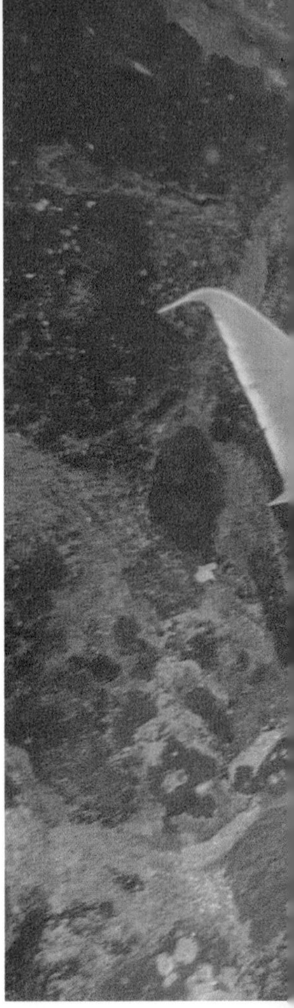

Lined Bristletooth *Ctenochaetus striatus*
Orangespine Unicornfish *Naso literatus*
Bluespine Unicornfish *Naso unicornis*
Sleek Unicornfish *Naso hexacanthus*
Spotted Unicornfish *Naso brevirostris*

Close to the reef-edge in the central Red Sea, the most common species is the beautifully iridescent blue striped surgeon – *Acanthurus sohal*. It is also perhaps the fish which is most likely to injure swimmers. However, the author has made numerous attempts to aggravate this species sufficiently for it to make a successful attack, but so far with little result.

ABOVE LEFT
Sailfin Surgeonfish
Zebrasoma veliferum
Photo F Bavendam

BELOW LEFT
Spotted Unicornfish
Naso brevirostris
Photo F J Jackson

Bluespine Unicornfish
Naso unicornis
Photo F J Jackson

"Sohal", as it is known in Arabia, is an agile fish which vigorously protects its "home range" from intruders – be they Man or fish. It does this by swimming rapidly towards the "trespasser" and then at the last moment curving away while at the same time thrusting its tail end towards its perceived aggressor. At the base of the tail, on each side of the caudal peduncle, there is a sharp "scalpel" which is capable of causing a slash wound in even the toughened skin of a hardy diver. Ninety nine times out of a hundred however the Surgeonfish just misses and no contact is made; but the sheer determination and perspicacity of its threatening behaviour is enough to deter one from remaining around long enough to see whether the Surgeonfish will strike on its next approach,

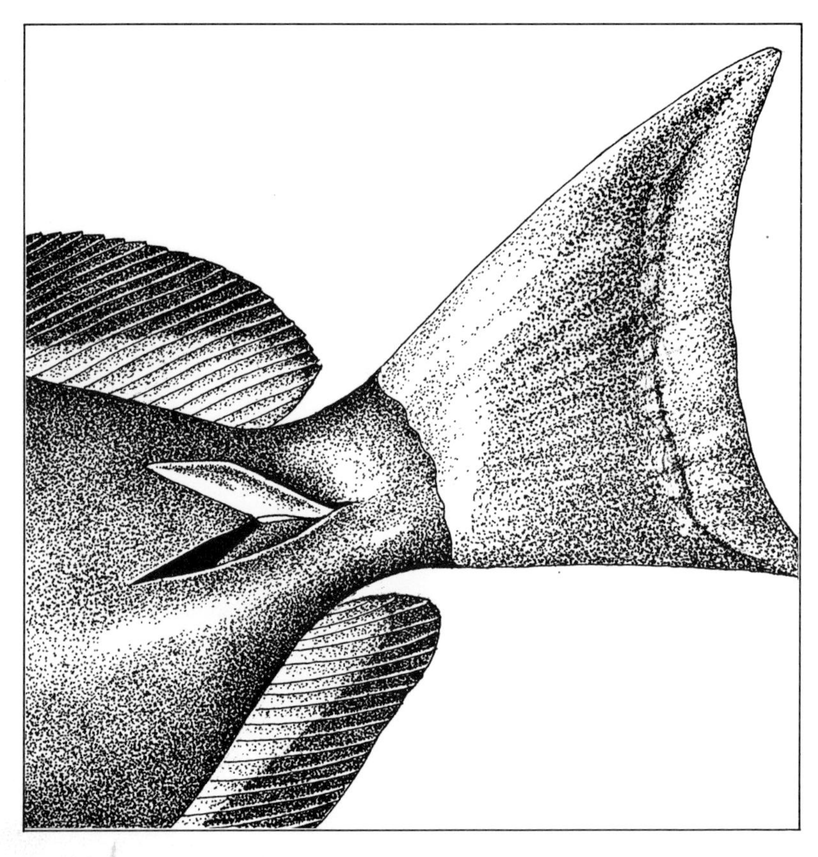

FIGURE 8
The scalpel apparatus of a Surgeonfish.

for it seldom leaves one in doubt that it intends to keep-up the attack until one has withdrawn to a respectable distance from its "home-range". Some of the other Surgeonfishes are less aggressive than "Sohal" but they still have the sharp scalpels on each side of their tail and they can make a deep cut in anyone who handles them carelessly. The majority of cuts from Surgeonfish are inflicted on inexperienced spearfishermen or on net and line fishermen.

There is a suggestion that Surgeonfish lacerations are also toxic since the degree of pain associated with the cut seems to be out of proportion with the size of the cut itself. This pain often spreads to the groin or axilla and it may last for several hours. Secondary infection is a frequent complication of such cuts.

First Aid treatment for Surgeonfish cuts involves the normal routine of laying the patient down with the affected region of the body in an elevated position. If there is little bleeding a very small incision parallel to the long axis of the limb will encourage bleeding and help to reduce pain. The wound should be washed, first with sea-water and then with hot water. If this relieves the pain, the wounded area should be immersed in hot water (50°C) for up to 30 minutes or until removal from the hot water is not accompanied by any worsening of the pain.

Additional medical treatment may include local anaesthesia (using an anaesthetic *without* adrenaline such as Lignocaine); tetanus prophylaxis; general symptomatic treatment (e.g. for cardiac shock or respiratory depression) and, if secondary infection occurs, use of antibiotics such as tetracycline (250mg q.i.d.). Preventive measures include the fairly obvious precaution against handling these fish in the sea. They have extracted their own reprisal on many spearfishermen!

Yellowtail Surgeonfish
Zebrasoma xanthurum
Photo F J Jackson

Black Surgeonfish
Acanthurus nigricans
Photo F J Jackson

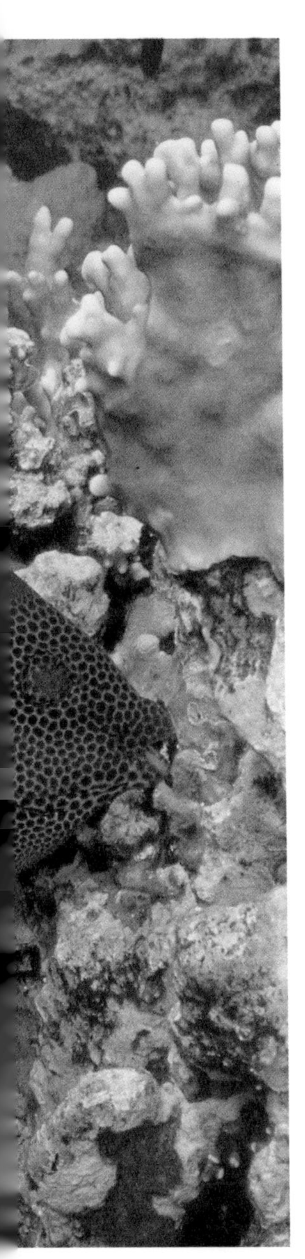

RABBITFISH *Siganidae*

There are four species of Rabbitfish in the Red Sea i.e. the Rivulated Rabbitfish (*Siganus rivulatus*), the Forktail Rabbitfish (*Siganus argenteus*), the Squaretail Rabbitfish (*Siganus luridus*) and the Stellate Rabbitfish (*Siganus stellatus*). They are all well known to fishermen and indeed the general public since "Sigan" are regarded as among the tastiest of Red Sea fish. Despite their culinary qualities however they have an impressive array of venomous spines – 13 dorsal, 4 pelvic and 7 anal. These spines have grooves extending along each side of the median line in which are situated – towards the distal portion – venom glands. The gland extends almost to the tip of the spine. When a spine punctures a victim's skin the venom is released and flows along the grooves into the puncture wound.

It is usually the first dorsal spine which causes stings since sudden movement of the fish drives this spine forward. The initial symptom of Rabbitfish stings is one of sharp and quite severe pain. This increases over the next few minutes and often becomes excruciating. The pain usually lasts for several hours but it may persist at less intensity for a few days. After the initially localised pain following the sting the pain may spread to regional lymph glands (usually groin or axilla) and the patient may become very distressed and mildly delirious. In a few cases signs of cardiovascular collapse have been reported including rapid pulse, hypotension and fainting. Respiratory distress can also occur in severe cases.

First Aid procedure follows the rest and elevation of affected region pattern previously described together with cleaning the wound and removal of foreign bodies; encouraging bleeding (by a small incision parallel to the long axis of the limb); immersion in hot water (50°C) and possible bathing of the wound with a weak solution of potassium permanganate.

Stellate Rabbitfish *Siganus stellatus* F J Jackson

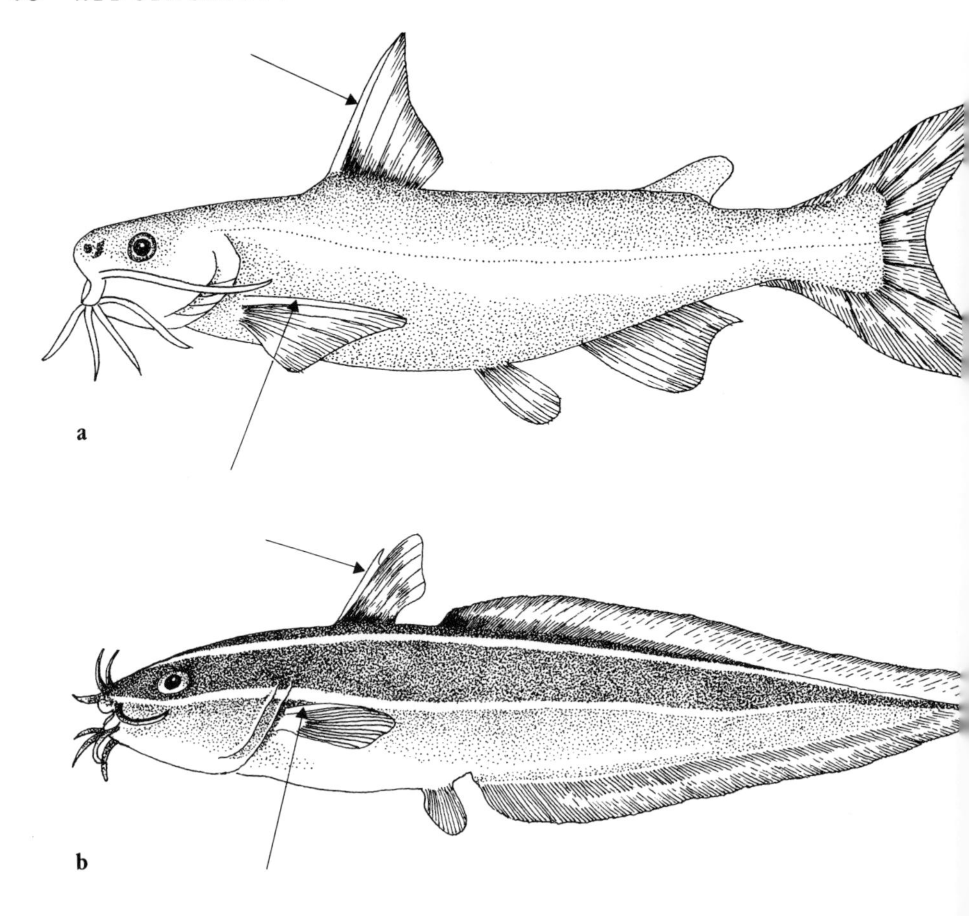

FIGURE 9
Red Sea Catfish and their poisonous spines.

a *Arius* sp. – poisonous spines marked by arrows.
b *Plotosus lineatus* – poisonous spines marked by arrows.

Additional medical treatment includes local anaesthesia (by an anaesthetic *not* containing adrenalin); symptomatic treatment of generalised symptoms (e.g. cardiac shock and respiratory depression); tetanus prophylaxis and if secondary infection occurs, use of an antibiotic such as tetracycline 250mgm q.i.d.

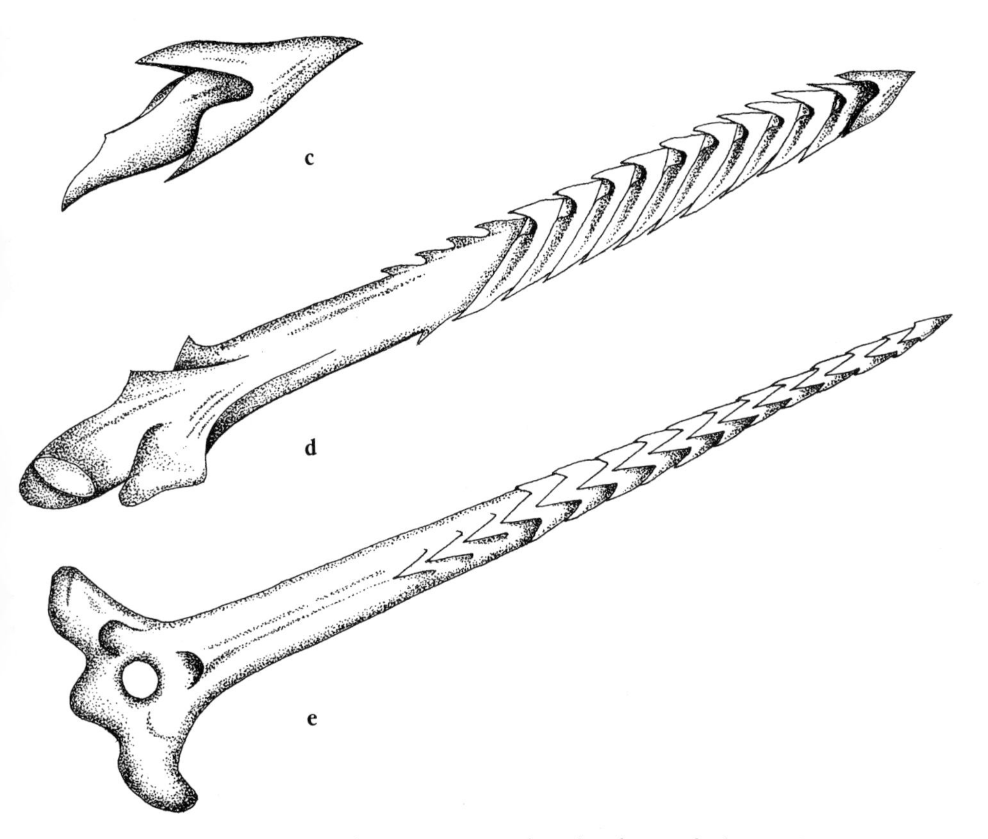

c Soft tip or spurious ray from dorsal spine of *P. lineatus*. Serrations along the whole spine are formed by a repetition of these spurious rays.
d and **e** Dorsal spine of *P. lineatus*.

CATFISH

Plotosus lineatus

These small black and white striped fish often swim in shoals and may be seen moving as a tightly knit group over areas of level sea-bed such as sea-grass meadows or sandy lagoons. They use their oral barbels to grovel in the sediment and to sense food items. They have single dorsal and pectoral spines which have retrorse dentations and are venomous (Fig. 9).

Striped Eel Catfish
Plotosus lineatus

Each spine is covered by a thin sheath of skin which contains venom glands. If a spine penetrates a victim's skin the thin covering is pierced and venom passes along the spine and into the wound. Despite their relatively small size (generally less than 30cm) they can inflict an extremely painful sting which has been reported on a few occasions to have been lethal.

The pain is said to be worst in the first ten minutes but it can remain sharp for an hour or so and may still be present as a dull throbbing sensation 48 hours after the sting occurred. The wound itself may take weeks to heal. One of the reported cases of stings from *Plotosus lineatus* involved a puncture wound in a thumb. The victim was unable to use the thumb for five and a half months!

Halstead, in his comments on poisonous catfish summarises the situation with regard to the Red Sea species of *Plotosus* as follows: "If the fish is *Plotosus lineatus*, a person would show even greater judgement by leaving it completely alone".

First Aid is similar to that described for most other venomous fish – i.e. rest; elevation of affected limb; cleaning of wound; immersion in hot (50°C) water; encourage bleeding; seek medical help. In addition, the wound can be bathed with sodium bicarbonate (baking soda).

Brief medical notes are given in Appendix 8. Preventive advice includes taking special care when handling fish nets – use of shoes when walking through shallow water together with the foot shuffle technique already recommended as a means of avoiding sting rays.

In addition to *Plotosus*, there is another Red Sea catfish, i.e. *Arius* sp. which occurs in deeper water and is more likely to be encountered by fishermen than by divers or reef-walkers. First aid and medical advice is the same as that for *Plotosus* described above and in Appendix 8.

While carrying out marine research in the Indian Ocean some years ago the author was diving in an area of relatively open sea, in a depth of about 20 metres, between two islands. The boat was manned by a local seaman who had considerable experience of the area. On this occasion he had been more than usually nervous during diving preparations and was visibly delighted upon my safe return. He later explained the circumstances in which a friend of his, diving in the same region, had recently lost a leg to a large Barracuda. The story is such a classic example of how Barracudas and Sharks can be invoked to attack swimmers that it is worth repeating.

His friend had been spearing fish along the reef-edge and was in the process of returning, with a *Caranx* on his spear, to the moored boat, when he caught site of a large solitary Barracuda (*Sphyraena barracuda*). Not wanting to give up his catch to this sleek predator he increased his pace towards the boat and as he reached it he raised the speared *Caranx* out of the sea to pass it to his colleague (the author's boatman). As soon as the spear-gun was out of his hands he started to hoist himself from the water and at that moment – with fish gone and body half out of the water – the Barracuda struck. One leg was neatly severed above the knee! The victim only survived as a result of immediate first aid and a rapid trip to the hospital.

The moral of the story is that spear-fishing is not only a destructive sport but can also be a dangerous one. There are many cases of attacks on divers caused by Barracudas or sharks being first of all excited by the vibrations and blood of dying fish and then being confused by a removal of the fish from the sea and finally, at the last moment, turning their attack on the diver. As one who disapproves of sport spearfishing on coral reefs it is a pleasure to find such a compelling argument against killing reef-fish with spear-guns. There is no doubt that it greatly increases the likelihood of attack by large reef predators. A secondary lesson is that if one finds oneself in the position of holding a speared fish in the water, do not suddenly remove it when sharks or barracudas are present.

BARRACUDAS

Sphyraenidae

There are five species of Barracudas in the Red Sea – i.e. the Great Barracuda (*Sphyraena barracuda*), the Pickhandle Barracuda (*Sphyraena jello*), the Chevron Barracuda (*Sphyraena putnamiae*), the Blackfin Barracuda (*Sphyraena genie*) and the Yellowtail Barracuda (*Sphyraena flavicauda*). Of these only *Sphyraena barracuda* has been implicated in provoked attacks on Man. The account given above is typical of the circumstances under which they may be expected to attack divers. Their reputation tends to be worse than their real threat however since they have an impressive array of sharp teeth which they frequently display to divers at close quarters. They are attracted by divers's bubbles and by bright shiny objects such as demand valves and they often hover in the water a few metres from SCUBA divers or snorkellers. This can be quite an unnerving experience – even to experienced divers – especially when a Barracuda follows a swimmer for long distances. Normally such incidents are *not* accompanied by any truly aggressive behaviour on behalf of the Barracudas.

Barracuda bites consist of straight lacerations and are unlike the jagged gashes caused by sharks. There is often

Great Barracuda *Sphyraena barracuda* Photo P J Vine

profuse bleeding associated with deep cuts. The victim suffers shock which manifests itself in a general pallor and profuse perspiration accompanied by a thready pulse, hypotension and possible fainting.

First Aid is directed towards stopping bleeding if necessary by use of a tourniquet. Ideally, medical attention should be brought to the victim who should not be moved unless absolutely necessary. Additional medical notes are given in Appendix 9B.

RIGHT
Schools of Pickhandle Barracuda
Sphyraena jello
Photo J Duran

Pickhandle Barracuda
Sphyraena jello
Photo P Vine

SHARKS

There is some doubt regarding the presence of the Great White Shark in the Red Sea. According to Dr. J.E. Randall who has carried out intensive studies on Red Sea fish (see "Red Sea Reef Fishes", Immel, 1983) the Great White does not occur in the Red Sea. The well known diver, Dr. Hans Hass (who was probably the first person to undertake underwater research in the Red Sea with modern diving techniques) claims however to have been attacked by a Great White Shark at Sanganeb reef.

In his book, Conquest of the Underwater World (published by David & Charles in 1975), Hass writes as follows:

"On June 14th, 1950, I myself was attacked by a white shark near the landing stage of the Sanganeb reef, in the Red Sea, eleven miles north east of Port Sudan. This specimen was thirteen feet long and came steadily towards me, like a machine. I had been filming smaller shark and only saw the creature when it was very close to me. I was at the top of a vertical face of the reef, where I was crouching on projecting coral at a depth of about 50 feet. To lure sharks we had chopped fish into pieces and tossed them into the sea near the edge of the reef. When I noticed the creature, it was already too late to reach for my spear, which was floating above me, attached to my shoulder by a loop. I gave a scream, as loudly as I could – but this shark did not react, so I was left with only my hands with which to ward off the great head. It must be emphasised that the creature displayed not the slightest excitement. The light-grey colossus swam quite slowly in a straight line along the wall directly towards me – just as if I were any old chunk of meat it wanted to snap up. When the snout was right in front of me, I punched at the back of its mouth, in the region of its gills. It turned away and circled round. The unexpected movement directed against it had produced the reaction of flight. Then steadily and unswervingly it came at me again – no slower and no faster. In the meantime I had had time to reach for my spear, and I

ABOVE RIGHT
Whitetip Reef Shark
Triaenodon obesus
Photo Ron & Valerie Taylor

Whitetip Reef Shark
Triaenodon obesus
F J Jackson

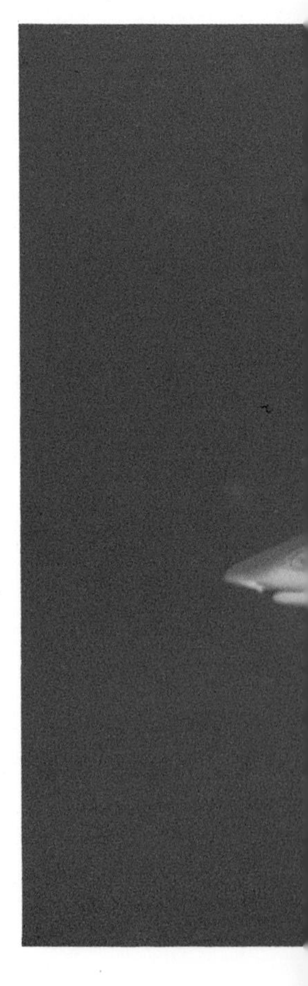

ABOVE AND RIGHT
Hammerhead sharks
Sphyrna mokarran
Photos F J Jackson

thrust the point at its head. It turned away. Now, however, I was attacked by the other shark, which I had been filming earlier, and which until then had been only curious but by no means aggressive. It was about ten feet long and wanted a piece of me too – an example of mood transference, of the infectiousness of predatory behaviour ...". Eventually, Hass extricated himself and escaped over the reef-crest and into the shallows.

The identification of the large Sanganeb shark as a Great White Shark is uncorroborated and while the description of its behaviour would perhaps tally with that of a Great White, there have been no other reliable reports of this species in the Red Sea. On balance, it seems likely that Hans Hass' early encounter was with one of the other large sharks which frequent deep waters around Sanganeb reef.

A detailed account of Red Sea sharks, written by Dr. J.E. Randall, is also being published by Immel, concurrently with this book. I do not therefore intend to enter into a lengthy discussion about the different species of shark which are present in the region. Suffice it to mention here that the most dangerous locally occurring species which have been implicated in various attacks on Man are as follows:

Tiger Shark *Galeocerdo cuvier*
Oceanic Whitetip *Carcharhinus longimanis*
Silvertip Reef Shark *Carcharhinus albimarginatus*
Hammerhead Shark *Sphyrna mokarran*
Hammerhead Shark *Sphyrna diplana*

Of these species the most familiar one to Red Sea divers is the Silvertip Reef Shark which lives close to deep water along the outer-edge of barrier reefs or around offshore pinnacle reefs. In his book "Shark", Philippe Cousteau describes an attack by *C. albimarginatus* in the Red Sea. The attack was stimulated by his colleague Raymond Kientzy spearing a *Caranx*. When the shark hit Raymond's SCUBA tank it turned and swam rapidly away. It is reported to have made unprovoked attacks on divers. *C. longimanis* tends to remain in open water away from reefs and divers rarely see it or Tiger sharks, which in the Red Sea seem to generally stay in deep water. Hammerhead Sharks may be seen at certain locations such as off Sanganeb reef where large numbers of them have been frequently observed by divers.

The sharks most frequently seen by divers are however less dangerous than these species but they may also be less shy of approaching people in the water. This may seem to be a contradiction but it is a fact that many of the large and potentially dangerous Red Sea sharks are also quite nervous in the initial stages of their approach to divers. In his book "Red Sea Reef Fishes", Dr. J.E. Randall illustrates what he regards as the four sharks most often encountered on Red Sea reefs. They are as follows:

Blacktip Reef Shark *Carcharhinus melanopterus*
Shortnose Blacktail Shark *C. wheeleri*
Sandbar Shark *C. plumbeus*
Whitetip Reef Shark *Triaenodon obesus*

Blacktip Reef Shark
Carcharhinus melanopterus
Photo S Hall

Of these, the Shortnose Blacktail Shark and Whitetip Reef shark are most likely to give cause for concern by their aggressive behaviour. In cases where fish are speared, the Whitetip Reef Shark is often the first shark to attack and there have been occasions when it has injured spearfishermen who have been carrying bleeding fish. Divers also record frequent meetings with the Shortnose Blacktip Shark which is closely related to the Grey Reef Shark of the Indo-Pacific regarded as an aggressive species and a menace to divers. Of the other two species in Dr. Randall's list, the Sandbar Shark is not known to have attacked Man while there have been some attacks by Blacktip Reef Sharks, mainly upon people wading in the shallows. Attacks by this species were analysed by J.E. Randall in a paper in Pacific Science (1973). He described ten incidents regarding this species and all but one involved people wading in shallow water. The sharks ranged from 18 inches to 5 feet long.

The question of why sharks attack divers or swimmers at the surface is an interesting one since the answer is not quite as obvious as it may seem. In general, if sharks behave aggressively towards Man, the chances are that they are stimulated by more than hunger. Territoriality may have more to do with aggressive behaviour than a desire to eat people! Many people who have encountered Red Sea sharks while diving will be familiar with aggressive displays which serve to frighten intruders away rather than being a definite prelude to an attack. Such displays often consist of:

1 snout lifting

2 lowering of pectoral fins

3 back arching

4 bending of the body laterally.

In most cases if the response to such a clear warning is to withdraw from the area, the shark does not pursue unless some additional factor has stimulated it. A shark's territory may consist of general 'no-go' area around its living space, wherever it may be – rather than a fixed territory. Such a warning display would normally be used towards other sharks and as with other territorial animals it is likely that, following a confrontation, one of the sharks would rapidly withdraw before battle commences. This insight may help to explain two facts which puzzle many divers – i.e. that big and potentially dangerous sharks also seem quite shy in their approach to strange objects (such as divers) and that they may often (but *not* always) be scared away by a sudden shout or movement underwater. In the give and take of the shark's world there are probably priorities which might run something like this:

1 Keep a clear living space around me and warn any intruders to respect this.

2 Avoid unnecessary conflict – i.e. if intruder does not withdraw then perhaps he is stronger than me so give *him* space and withdraw myself.

3 If the intruder does *not* withdraw but is obviously weaker than me – check it out again and if necessary give it a bite.

4 If the bite tastes nice have some more!

Despite their tremendous power and the potential danger which sharks pose to Man, in their own world they have competitors and enemies in the form of other sharks and dolphins. Their behavioural patterns help to keep the peace among sharks. When a diver retreats from an aggressive shark the stimulus which caused the display of anger is gone and peace reigns once more. If instead of retreating a diver suddenly rushes towards a shark and makes noise underwater, the shark may follow the second rule in its book: i.e. avoid unnecessary conflict; and then withdraw itself. There are however circumstances when this approach does not work and by far the safest procedure is to retreat gracefully. The author can recall many meetings with Red Sea sharks which fit this general pattern and a few incidents which fall completely outside it. The latter category includes what one might recognise as true attacks rather than warning displays. Such attacks have nearly always been brought on by spearfishing and by the attraction of injured fish.

A recent study on shark attack behaviour has been carried out by Ron and Valerie Taylor, together with Jeremiah Sullivan. The study involved use of a chain-mail armour suit and Valerie Taylor as the bait! A brief review of their results has been published in National Geographic Magazine (May, 1981) and in a booklet entitled "The Great Shark Suit Experiment" (Ron and Valerie Taylor, 1981). The Taylors must be regarded as among the foremost authorities on shark-diver encounters and the shark-suit experiments were much more significant than may appear at first sight. Apart from a number of truly dramatic photographs featuring Valerie's arm in various shark jaws, this husband and wife team observed and recorded the attacking behaviour of different sharks. With reference to species occurring in the Red Sea their comments were as follows:

Whitetip Reef Sharks – *Triaenodon obesus* are often the first to appear and are relatively fearless of divers. When they attack, their first bite is "their best" and subsequent ones tend to diminish. It is important for them to get a good grip with their bottom holding teeth. They release their grip by retracting their teeth into their gums, opening their jaws and then backing-off. The real cutting action comes from their

Whitetip Reef Shark
Triaenodon obesus
Photo Ron & Valerie Taylor

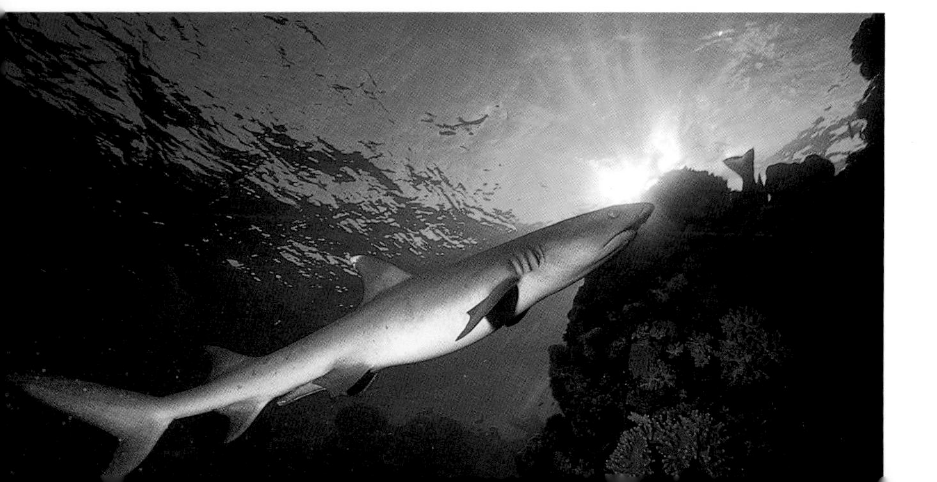

Whitetip Reef Shark and Embryos
Triaenodon obesus
Photo Ron & Valerie Taylor

top teeth when the white-tip shakes its head back and forth. They often brace themselves against the coral in order to maximise their leverage while sawing through their prey with their upper teeth.

Oceanic Whitetips *(Carcharhinus longimanis)*
These sharks tend to bump an intended prey before attacking. If after one or two bumps there is no retaliation from the prey, they then bite and use the characteristic sawing motion. Valerie Taylor points out that Oceanic Whitetips are deceptive creatures because they tend to move quite slowly and calmly and may leave a diver with the impression that there is not much to fear and then, with no visible change of pace, the shark may switch from bumping to biting.

Tiger Sharks: in my experience of diving in the Red Sea, I have rarely seen Tiger Sharks in shallow water although I am well aware that they may hunt their prey in the shallows. On one occasion while snorkelling through waist-deep water along Australia's northern Great Barrier Reef – I was calmly followed by a large tiger shark with its dorsal fin completely out of the water. In the Red Sea however the tiger sharks which I have encountered have been caught in trawlnets which had been fishing at about 600 feet deep – off Farasan Bank.

Valerie Taylor describes their behaviour as "confident around baits" but she also claims that they are relatively easy to work with underwater and says that she and Ron Taylor have never experienced unprovoked attacks from them.

Valerie Taylor makes the point that "potentially dangerous sharks are on the whole not very keen to bite a human". This observation, which is based upon a great deal of experience, confirms my own impression gained after numerous encounters with Red Sea and Indo-Pacific sharks. Filming reef-sharks underwater can be a quite frustrating experience since they are often very shy of divers and the sound of a cine camera motor may cause them to turn back from the most appetising bait. Of course, once they become excited they lose their fear and may become more aggressive.

A final footnote on the subject of shark behaviour:
In certain cases the frontal approach used to scare away sharks may actually stimulate them to attack rather than discourage them. I repeat that the best defence when threatened is calm retreat!

If a shark attack does occur (and you just happen to have this book available for reference!) the following first aid instructions should be followed:

1 Remember that shark victims usually die from blood loss and shock and the immediate aim of first-aid treatment is to reduce these two factors.

2 Remove victim from water and place so that legs and trunk are elevated slightly above head. This will increase blood flow to the head and help to reduce the effects of shock.

3 Minimise bleeding by pressing on pressure points or if necessary applying a tourniquet. You may start this process even before the victim is removed from the sea.

4 Do *not* move the victim further than necessary – i.e. just take out of water and start treating immediately – do *not* try to make a mad mercy mission to hospital but summon medical aid to come to the patient. There are many cases where shark attack victims might have survived if they had been given on-the-spot first-aid instead of being rushed to hospital.

As a general rule, the patient should not be moved for at least 30 minutes regardless of whether he is receiving treatment.

5 Do *not* give any warm drinks or alcohol. Only a little fresh water should be given for the patient to sip.

Ideally, if the bleeding is thus stopped and the patient is left still, kept warm, and is given plenty of reassurance, the first-aid stage is complete and all subsequent aid should be administered by a doctor.

Some notes on medical treatment of shark attack victims are given in Appendix 9A.

The main preventive suggestions for divers and swimmers are as follows:

1 Do not swim with bleeding wounds.
2 Do not urinate in the water.
3 Do not carry fish or spear fish underwater.
4 Do not test a shark's patience – if it displays aggressively, withdraw cautiously.
5 Move steadily without panic.
6 Carry a stick or shark defence mechanism.
Do not dive alone.
7 Wear some form of protective clothing.

The experienced diver, researcher and film-maker Walter Starck has successfully tested a black and white (horizontally) striped wet-suit as a means of confusing visual recognition by sharks. Although there is some controversy regarding how effective this is, there is little doubt that in many situations it gives a diver an advantage over those who have plain black wetsuits. This is enough by itself to justify more widespread use of striped wetsuits by those diver-photographers who seek to commune with Red Sea sharks.

MORAY EELS

Muraenidae

Moray Eels are not normally aggressive towards Man. Injuries caused by them are not unknown however since the eels frequently hide in crevices and remain concealed from divers who reach into these holes. Several particularly bad injuries attributed to Morays have been caused by people sitting on jetties or boats and dangling their legs in the sea. If they happen to be close to a Moray Eel's "den" then there is a possibility that the eel will consider the person's feet as attractive bait.

The reputation of Morays has been discussed by Randall (1969) in a paper entitled "How dangerous is the Moray eel?" they do not fully deserve their loathsome reputation of deadly beasts to be avoided at all costs and killed at every opportunity. In a bid to lay the myth of their unpleasant nature, various divers have tamed Moray Eels by regularly feeding them. Valerie Taylor has two favourite Morays known as Harry and Fang whom she feeds and caresses, and Jeddah-based divers Georg and Monica Jungbauer have made a film on Red Sea Morays and their unfortunate plight at the hands of ignorant spearfishermen. There is perhaps a danger in this that less experienced divers will be lulled into believing that Morays pose no threat whatsoever.

Moray bites are particularly painful and can be dangerous. Secondary infection is a regular and serious complication and inadequately treated wounds may become gangrenous. Tetanus may also result if prophylaxis is not administered.

First Aid should be immediately directed at stopping bleeding by local pressure and if necessary a tourniquet. Medical attention should, if possible, be brought to the victim. Additional medical notes are given in Appendix 10.

Since Moray Eels are generally poisonous to eat (Khlentzos, 1950) there is no justification for hunting them. Leave them alone and do not aggravate them. If contact with them is inevitable wear heavy protective clothing.

Yellowmouth Moray
Gymnothorax nudivomer
Photo J E Randall

GENERAL FIRST AID IN THE MARINE ENVIRONMENT

Most accidents and injuries among Red Sea divers, snorkellers, swimmers, reef-walkers, fishermen or boating enthusiasts are likely to occur at some considerable distance from medical facilities and in many cases there is an inevitable long delay between an incident and the patient arriving at hospital or a doctor's surgery. In virtually all cases discussed in this book, rapid first aid greatly enhances the likelihood of a successful recovery. In some situations it will make the difference between the patient living or dying.

In many of the potential injuries which have been discussed there are standard first-aid procedures which should be applied. Some of these are described in more detail in this chapter. In addition, there are two other medical problems which frequently occur among those who dive on Red Sea reefs. These are coral-cuts which develop secondary infections and turn into abscesses; and "coral-ear" which is a painful fungal infection caused by spending long periods in the sea and living in hot, humid conditions which promote fungal growth. These two ailments are also discussed below.

I. ARTIFICIAL VENTILATION

The modern view is that the most effective means of providing artificial ventilation is the Expired Air Resuscitation technique or E.A.R. for short. The advantages of E.A.R. over other methods are as follows:

1 It provides more effective pulmonary ventilation than other first aid systems.
2 It is easy to learn and practice.
3 It is simple to carry out.
4 It will not cause injury.
5 It can be started while the person is still in the sea.
6 It can be done by one person.
7 The rescuer has both hands free.
8 No aids are required.
9 The casualty remains on his or her back and in a suitable position for cardiac massage to be given – should this be necessary.

The method is summarised as follows:

1 Place casualty on his back with head slightly raised and clear debris from nasal and oral airways.

2 Lift the jaw and tilt the head back by pressure on the chin as far as it is reasonably possible (Fig. 10).

3 If breathing starts, continue close observation and turn the casualty over onto his front, making sure that the air passages are kept clear.

4 In cases where severe injuries have occurred however the patient should be left lying on his back and moved as little as possible

5 If breathing does not start the mouth should be cleared of any debris. The jaw lift and head tilt described in (1) above should be maintained. With the casualty's mouth closed, breath steadily into the nose.

An eye should be kept on the casualty's chest to check that it has inflated. If it has not done so, extend the neck and lift the jaw a little more so that the airway passage is straightened out and repeat blowing. If the chest does still does not inflate switch to a mouth-to-mouth technique – i.e. inflating through the mouth and holding the nose closed.

6 The rhythm which should be established is as follows:
a Give four quick breaths to partially inflate chest.
b After that give one breath every five seconds – taking about 2 seconds to actually blow air into the casualty.
c This should be maintained until the casualty regains consciousness or until help arrives.
d If there are no serious injuries the casualty can be placed over onto his side – in a recovery position – once breathing is re-established.

Infant Resuscitation

With babies, the method is slightly different. Place your mouth over the nose and mouth of the baby and puff air into it from your cheeks rather than with deep breaths from your lungs. One puff every 2 or 3 seconds is about the right frequency of inflation.

Complications Due to Vomiting

It is quite possible that a casualty will vomit. It is essential that the person is turned onto the side so that vomit is expelled rather that inhaled. The mouth should be cleared before resuscitation recommences.

II. EXTERNAL CARDIAC COMPRESSION

Cessation of breathing usually causes the heart to stop beating so it will often be necessary to combine Expired Air Resuscitation described above with heart stimulation in the form of external cardiac compression.
The following signs indicate that the heart has stopped beating.

1 No pulse can be felt. Check the carotid artery in the neck.

2 Dilated pupils may result from lack of oxygen to the brain caused by cardiac arrest.

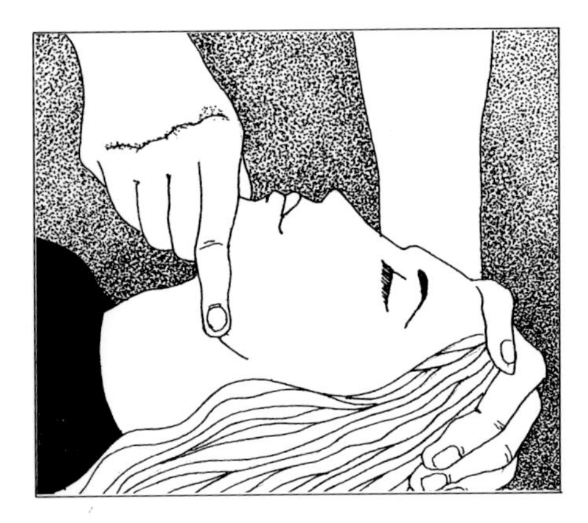

a Tilt head back and gently raise jaw. This may help to clear airways and sometimes stimulates breathing to recommence.

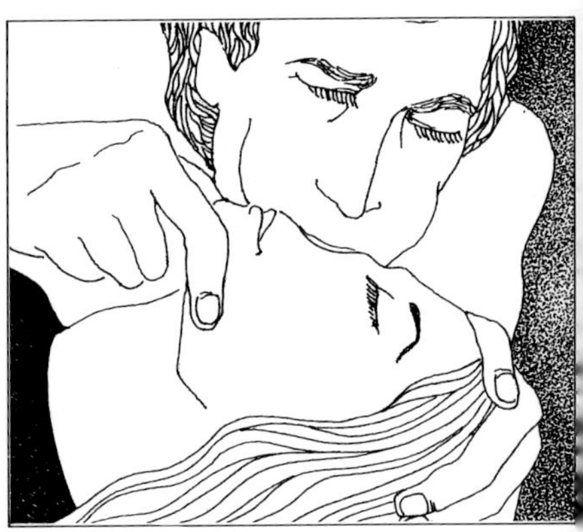

b If the breathing has not restarted clear any obstructions from mouth, maintain jaw lift and, keeping mouth closed breathe steadily into the nose. Make sure that the chest inflates. If it does not do so repeat after increasing the neck stretch and jaw lift. If it still fails to inflate change to mouth to mouth resuscitation.

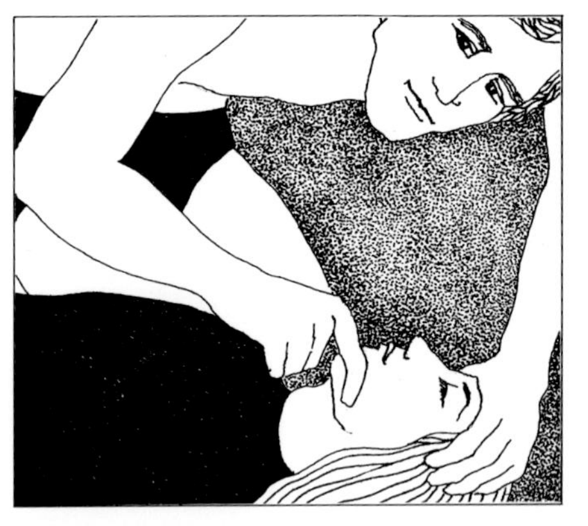

c In the case of mouth to nose resuscitation continuing; maintain a regular rhythm of inflation as follows: Initially give four breaths to inflate the chest, then give one breath every five seconds – taking about two seconds to actually blow and three seconds to recover and take a breath before the next inflation.

d In the event that mouth to nose resuscitation fails to inflate lungs switch to mouth to mouth resuscitation. This is carried out by holding the casualty's nose closed and supporting the head as illustrated in the neck extended, mouth open and jaw raised position.

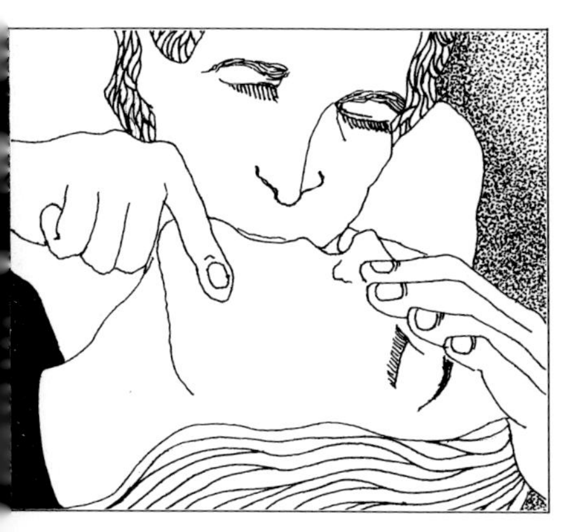

e Inflation is achieved by sealing your lips over those of the patient and blowing steadily. The basic sequence is to give four initial blows followed by one every five seconds (i.e. the same rhythm as described in 10c).

f In the case of small children or infants the technique should be modified to cover both the nose and mouth with your own mouth and to puff gently from the cheeks rather than using the full force of a blow from your own lungs. The rhythm is to give one puff every three seconds approximately.

FIGURE 10
Artificial Ventilation.

a, **b**, **c**, mouth to nose resuscitation

d, **e**, mouth to mouth resuscitation

f, infant or small child resuscitation

3 Skin and/or lips turn blue despite effective E.A.R.
As soon as the heart stops beating External Cardiac Compression (E.E.C) should begin. The method is as follows:

a Give at least five full cycles of E.A.R. before checking for cardiac arrest.
b Raise the feet slightly above the head. Keep up E.A.R.
c Kneel alongside the casualty and feel for the lower part of the breast bone. Using the heel of a clenched fist give this a single sharp smack. This may stimulate the heart to start up again so check the pulse again.
d If the heart has not re-started, commence regular E.C.C. and keep up E.A.R.

The combined rhythm of respiratory and cardiac resuscitation is as follows:

1 Give a full inflation of lungs.

2 Place heel of one hand on the lower half of the sternum, keeping fingers off the chest. Place the other hand on top and keep both arms straight (Fig. 11).

3 Lean forward so that arms are vertical and press down on the breast bone by applying only enough pressure to depress it by about 4m. If the casualty is a fully grown adult around 25 to 30kg is required to achieve this. After each compression, release the pressure and repeat at a rate of one compression per second.

In the case of young children (under 10 years old) sufficient pressure can be applied with one hand and a compression rate of 80 to 90 per minute is required with less pressure. If the casualty is a baby you only need to press with two fingers but the rate should be supported with the free hand.

4 Give five compressions followed by one ventilation in a continuous sequence. (Do not give compression and inflation at the same time).

5 Continue the combined ventilation and heart massage until a pulse beat is re-established and the patient is breathing steadily. It is a good plan to make a rapid check on the patient's condition and re-establishment of heart and lung functions every five minutes but do not stop the rhythm of ventilation and compression for long in order to do this.

6 In the event that heart and breathing do not re-start – keep up the resuscitation until medical help arrives and instructs otherwise.

If two people are available one should work on the ventilation of the casualty while the other concentrates on the heart massage. In this case the sequence is that the first operator gives one full inflation and then leans back – holding the neck in the extended position – while the second operator gives five cycles of cardiac compression. The sequence is them repeated.

FIGURE 11
Heart massage and combined breathing resuscitation.

a Diagram to show the position of sternum. The lower (shaded) portion should be struck once with a clenched fist at the commencement of heart resuscitation.

b Position of rescue worker at the commencement of heart massage. Breathing resuscitation must also be maintained and the recommended rhythm is to give five compressions of the sternum at a rate of one per second, followed by one inflation of the lungs.

c and **d** These two diagrams illustrate how two rescue workers can provide heart massage and breathing resuscitation. The two treatments should not be given simultaneously. The rhythm is that one operator gives a full inflation of the lungs and then leans back while the other worker gives compressions of the sternum. The sequence is then repeated.

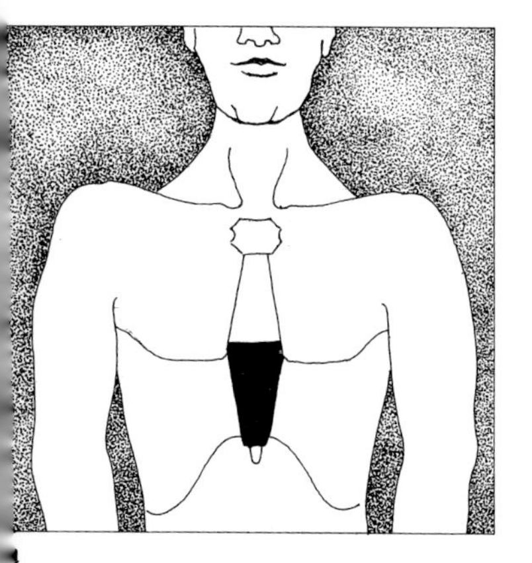

a

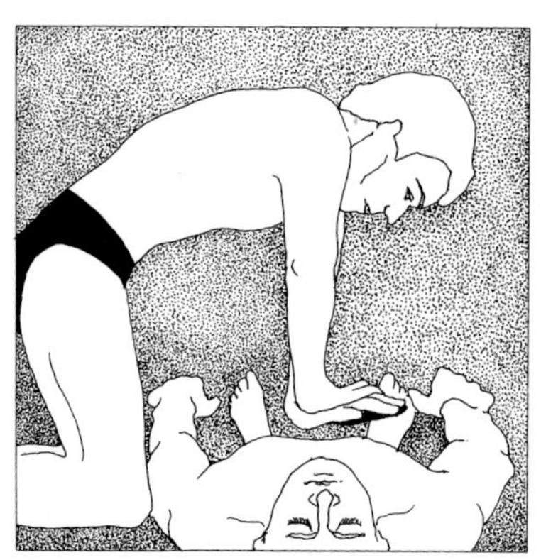

b

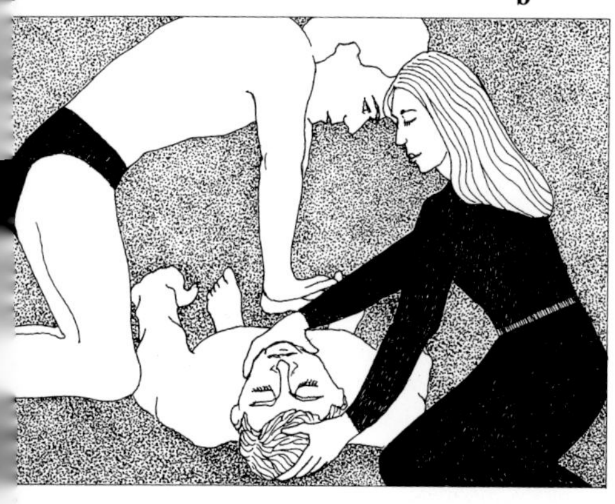

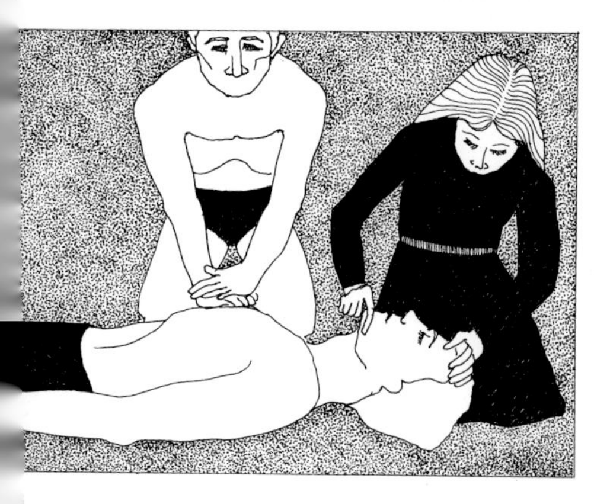

NB. It is dangerous to practice heart massage in training sessions since there is a real risk of causing injury to the sternum or ribs. In practice sessions the positions and movements can be demonstrated *without* applying any pressure.

One final word of warning about external cardiac compression is that practice sessions should not include applying pressure to the breast bone since this could cause injury.

III. CORAL CUTS

Many people who enter the Red Sea to explore its beautiful coral reefs suffer cuts or abrasions from sharp edged corals. Coral cuts are not in themselves dangerous but they do frequently develop into festering sores or ulcers. Such open wounds may take weeks or months to heal. If the cut occurs in areas of the body where there is a thin covering of skin and flesh over bones – such as on the front legs, below the knee, then healing may be extremely slow unless correct treatment is given.

The author can vouch for this fact, having suffered the effects of ulcerated coral cuts, which resisted healing for more than nine months. Twenty years after the event a large scar is still prominent. The moral is clear: do not under-estimate cuts caused by corals – even if they start as mere scratches. Follow first-aid instructions and take more care to avoid being cut in future.

Cuts caused by corals should be treated as follows:

1 Clean the wound and any scratches as soon as possible with an antiseptic lotion such as merurochrome, hydrogen peroxide or surgical spirit.

2 Apply an antibiotic powder as soon as possible.

3 Keep the wound(s) dry and covered by a dressing.

4 If abcess formation occurs a broad spectrum antibiotic such as tetracycline 250mg q.i.d. should be administered.

5 In cases where healing does not progress, an X-ray may reveal foreign bodies remaining in the surrounding tissues.

6 Tetanus prophylaxis is standard procedure.

Coral cuts can be prevented by wearing protective clothing while swimming on coral reefs.

IV. CORAL EAR

This can become a serious problem for divers in the Red Sea if they are spending long periods underwater and living in humid conditions so that their ear tubes do not dry out properly. Intensive diving such as occurs on diving holidays or marine scientific expeditions can lead to many members of a diving team being laid-off by ear-ache.

It can be prevented, to some extent, by the following precautions.

1 Wash ears in fresh water after diving.

2 Dry ears thoroughly after rinsing them out.

3 If faint ear-ache persists continue the fungicidal ear-drops as an immediate treatment and prophylaxis.

4 If the ear-ache persists, continue the fungicidal treatment, stop diving and stay out of the sea until the ear-ache has gone and the ears have remained free from pain for at least three days.

5 If no relief is provided by ear drops take medical advice which may propose use of a broad spectrum antibiotic (e.g. tetracycline 250mgm q.i.d.).

"Coral-ear" is less likely to occur among people who dive once a week or just at week-ends. It is also inhibited by the cool dry air created by air-conditioning systems.

Extreme pain is often associated with "coral-ear" and effective analgesics will be required. In severe cases morphine has been used.

Board Sailing Safety

Board sailing is one of the most exciting and energetic ways of enjoying the Red Sea. It combines the thrills of fast sailing with the enjoyment of cooling-off by occasional unplanned swims. Learning to wind-surf can be a haphazard affair in which the pupil is left to his or her own devices with a few words of initial advice and much subsequent encouragement from one's friends. In order to avoid accidents there are a number of rules which should be followed and these apply as much to the novice as to those experienced board-sailors who may feel that they are above taking such precautions.

1 Learn the techniques at a properly organised training school or under expert personal tuition. It is worth hiring a board and trying several makes before selecting one which suits you. Begin with a small sail and short boom since this will make the rig much easier to handle. The best place to practice is on an enclosed or semi-enclosed body of water – preferably on days when these are not too crowded by other water users. Always wear a buoyancy aid and select a sail with a large window so that you can have a clear view in all directions. Take care to ensure that the board is properly rigged and that the base of the mast is attached to the board.

2 Protect yourself against the weather. Even in the Red Sea one can suffer from exhaustion due to cold. Wind evaporation in winter can really cause one to become chilled.

Wet suits (including booties) are generally a good protection against this form of heat loss exhaustion and it is important to select comfortable wet-suits which one will not be tempted to leave ashore. In strong winter winds it is worth wearing a light windproof jacket over the wet-suit. Wear a hat and use sun-block cream to avoid sun-burn. As soon as you start to feel tired return to shore and take a rest.

3 Beginners should not sail too far from land. One should try to sail on close reaches along the shore, back and forth, keeping parallel with the land. If wind conditions prevent this, sail for a short distance offshore and then return, never venturing further out than you feel confident of paddling back should the need arise.

4 ***Maintain your equipment in good condition***. The board should have a towing eye and a safety line which links the rig to the board. Check the mast pivot and ensure that the sail is properly strung and in good condition. A quick release harness is worth using to reduce fatigue. Ideally this should be fitted with a back-pack in which can be stored waterproof emergency flares.

5 ***If you require assistance***, do not leave the board. It is a golden safety rule that one should remain with the board and not be tempted to leave it in order to swim ashore. Kneel or sit on it and wave one arm slowly from side to side. This is the board sailor's sign for "help needed". Save the emergency flares as a last resort – for when you are in imminent danger. If you are in difficulty and unable to sail back to shore carry out the self-rescue technique *before* it is too late. This is done as follows:-

a Remove the boom and roll the sail neatly around the mast – tie it down with an out-haul. Place the rolled sail and boom under you, stowed lengthways on the board.

b Take out the centre-board and either stow it with the rest of the rig or use it as a paddle. Lie on top of the board and paddle to shore using hands or the centre-board.

c Head towards the nearest land or anchored object where you can rest without drifting further offshore.

d If paddling is difficult ditch the rig but do not leave the board.

Board Sailor's Safety Code

i Do not sail alone and never at night. Ideally sail in a group and help each other.

ii Wear buoyancy aids.

iii Wear clothing to keep you warm and protect you from sun-burn.

iv Make sure a friend, colleague or official knows where you are going and always check back with the same person when you return.

v Listen to the local weather forecast and avoid sailing in strong winds unless you are very experienced and fully able to manage.

vi Do not participate in rough weather sailing unless first-class rescue facilities are available and dedicated solely to taking care of participating board sailors.

vii Carry a smoke-flare and basic rations.

viii Keep clear of boats, whether moored or under way. Take care to avoid divers, swimmers and children. Give full consideration to other water users.

ix Keep the shore in sight.

USE OF DECOMPRESSION TABLES

One of the most common accidents to occur among those who dive in the Red Sea is decompression sickness. When diving conditions are so ideal it is difficult to remember to check dive-times and to take necessary decompression precautions. Many divers tend to shy away from decompression tables in the mistaken view that they are immune from the danger of "bends" if they only use a single cylinder. On other occasions experienced divers are so completely entranced with the underwater environment that they put on a second cylinder and re-enter the sea without giving a moment's thought to "residual nitrogen times". It is absolutely essential that divers do pay heed to the danger of "bends" and that they plan their dives with decompression tables. Diver carried decompression meters are also worth using providing not too great a reliance is placed upon them.

The tables which are reproduced on pages 125-130 are based upon US Navy Diving Manual (1970) and they provide information on how to calculate the required decompression stops on a single dive as well as the "residual nitrogen time" which one must allow on subsequent dives. They are quite easy to use providing one follows the correct sequence of steps which are explained below.

1 Use table 1 to select the depth to which you will dive and the time spent at the bottom. For example: Depth = 18m; Bottom time = 100 minutes. Alternatively use table 2A to select a dive depth and duration which is within the "no decompression limits".

2 Read along the column of table 1 horizontally to ascertain:

a How long should elapse between leaving the sea bed and making the first decompression stop; e.g. in our example 50 secs.

b The duration and depth of the required stop(s); e.g. in our example: 14 minutes at 3 metres.

c The total ascent time (15 minutes in the quoted example).

d The "Repetitive dive group" which will be a letter (e.g. 'M' in our example).

3 This information allows you to plan a single dive. Thus, if one is to spend 1 hour and 40 minutes at 18 metres, a 14 minute decompression stop at 3 metres will be required. In addition it will store-up the information on that dive so that you can apply it to a future dive.

4 The safest way for amateur divers to use the tables is to plan dives within the "no decompression limits"; i.e. so that no decompression is required. For example up to 200 minutes at 12 metres; or 100 minutes at 15 metres or one hour at 18 metres. While table 1 does not give information on dives which fall within the no decompression limits, table 2A does provide information on this in order that the effects of a "no decompression dive" can be taken into account on a repeat dive. It must be remembered that while a single dive within "no decompression limits" does not require a decompression stop, if one re-enters the water for a second dive within twelve hours there is a residual effect from the previous dive which must be taken into account when calculating the decompression requirements of the second dive.

Standard Air Decompression Tables based on US Navy Diving Manual, 1970

DEPTH METRES	BOTTOM (TIME) MINS	TIME TO FIRST STOP (MIN:SEC)	DECOMPRESSION STOPS					TOTAL ASCENT (MIN:SEC)	REPETITIVE GROUP
			M 15	12	9	6	3		
12	200	–					0	0:40	•
M.	210	0:30					2	2:40	N
	230	0:30					7	7:40	N
	250	0:30					11	11:40	O
	270	0:30					15	15:40	O
	300	0:30					19	19:40	Z
15	100	–					0	0:50	•
M.	110	0:40					3	3:50	L
	120	0:40					5	5:50	M
	140	0:40					10	10:50	M
	160	0:40					21	21:50	N
	180	0:40					29	29:50	O
	200	0:40					35	35:50	O
	220	0:40					40	40:50	Z
	240	0:40					47	47.50	Z
18	60	–					0	1:00	•
M.	70	0:50					2	3:00	K
	80	0:50					7	8:00	L
	100	0:50					14	15:00	M
	120	0:50					26	27:00	N
	140	0:50					39	40:00	O
	160	0:50					48	49:00	Z
	180	0:50					56	57:00	Z
	200	0:40				1	69	71:00	Z
21	50	–					0	1:10	•
M.	60	1:00					8	9:10	K
	70	1:00					14	15:10	L
	80	1:00					18	19:10	M
	90	1:00					23	24:10	N
	100	1:00					33	34:10	N
	110	0:50				2	41	44:10	O
	120	0:50				4	47	52:10	O
	130	0:50				6	52	59:10	O
	140	0:50				8	56	65:10	Z
	150	0:50				9	61	71:10	Z
	160	0:50				13	72	86:10	Z
	170	0:50				19	79	99:10	Z
24	40	–					0	1:20	•
M.	50	1:10					10	11:20	K
	60	1:10					17	18:20	L
	70	1:10					23	24:20	M
	80	1:00				2	31	34:20	N
	90	1:00				7	39	47:20	N
	100	1:00				11	46	58.20	O
	110	1:00				13	53	67:20	O
	120	1:00				17	56	74:20	Z
	130	1:00				19	63	83:20	Z
	140	1:00				26	69	96:20	Z
	150	1:00				32	77	110:20	Z

DEPTH METRES	BOTTOM (TIME) MINS	TIME TO FIRST STOP (MIN:SEC)	DECOMPRESSION STOPS					TOTAL ASCENT (MIN:SEC)	REPETITIVE GROUP
			M 15	12	9	6	3		
27	30	–					0	1:30	•
M.	40	1:20					7	8:30	J
	50	1:20					18	19:30	L
	60	1:20					25	26:30	M
	70	1:10				7	30	38:30	N
	80	1:10				13	40	54:30	N
	90	1:10				18	48	67:30	O
	100	1:10				21	54	76:30	Z
	110	1:10				24	61	86:30	Z
	90	1:00				32	68	101:30	Z
	130	1:00			5	36	74	116:30	Z
30	25	–					0	1:40	•
M.	30	1:30					3	4:40	I
	40	1:30					15	16:40	K
	50	1:20				2	24	27:40	L
	60	1:20				9	28	38:40	N
	70	1:20				17	39	57:40	O
	80	1:20				23	48	72:40	O
	90	1:10			3	23	57	84:40	Z
	100	1:10			7	23	66	97:40	Z
	110	1:10			10	34	72	117:40	Z
	120	1:10			12	41	78	132:40	Z
33	20	–					0	1:50	•
M.	25	1:40					3	4:50	H
	30	1:40					7	8:50	J
	40	1:30				2	21	24:50	L
	50	1:30				8	26	35:50	M
	60	1:30				18	36	55:50	N
	70	1:20			1	23	48	73:50	O
	80	1:20			7	23	57	88:50	Z
	90	1:20			12	30	64	107:50	Z
	100	1:20			15	37	72	125:50	Z
36	15	–					0	2:00	•
M.	20	1:50					2	4:00	H
	25	1:50					6	8:00	I
	30	1:50					14	16:00	J
	40	1:40				5	25	32:00	L
	50	1:40				15	31	48:00	N
	60	1:30			2	22	45	71:00	O
	70	1:30			9	23	55	89:00	O
	80	1:30			15	27	63	107:00	Z
	90	1:30			19	37	74	132:00	Z
	100	1:30			23	45	80	150:00	Z
39	10	–					0	2:10	•
M.	15	2:00					1	3:10	F
	20	2:00					4	6:10	H
	25	2:00					10	12:10	J
	30	1:50				3	18	23:10	M
	40	1:50				10	25	37:10	N
	50	1:40			3	21	37	63:10	O
	60	1:40			9	23	52	86:10	Z
	70	1:40			16	24	61	103:10	Z
	80	1:30		3	19	35	72	131:10	Z
	90	1:30		8	19	45	80	154:10	Z

Depth Metres	Bottom (Time) Mins	Time to First Stop (Min:Sec)	Decompression Stops						Total Ascent (Min:Sec)	Repetitive Group
			M	15	12	9	6	3		
42	10	–						0	2:20	•
M.	15	2:10						2	4:20	G
	20	2:10						6	8:20	I
	25	2:00					2	14	18:20	J
	30	2:00					5	21	28:20	K
	40	1:50				2	16	26	46:20	N
	50	1:50				6	24	44	76:20	O
	60	1:50				16	23	56	97:20	Z
	70	1:40			4	19	32	68	125:20	Z
	80	1:40			10	23	41	79	155:20	Z
45	5	–						0	2:30	C
M.	10	2:20						1	3:30	E
	15	2:20						3	5:30	G
	20	2:10					2	7	11:30	H
	25	2:10					4	17	23:30	K
	30	2:10					8	24	34:30	L
	40	2:00				5	19	33	59:30	N
	50	2:00				12	23	51	88:30	O
	60	1:50			3	19	26	62	112:30	Z
	70	1:50			11	19	39	75	146:30	Z
	80	1:40		1	17	19	50	84	173:30	Z
48	5	–						0	2:40	D
M.	10	2:30						1	3:40	F
	15	2:20					1	4	7:40	H
	20	2:20					3	11	16:40	J
	25	2:20					7	20	29:40	K
	30	2:10				2	11	25	40:40	M
	40	2:10				7	23	39	71:40	N
	50	2:00			2	16	23	55	98:40	Z
	60	2:00			9	19	33	69	132:40	Z
	70	1:50		1	17	22	44	80	166:40	Z
51	5	–						0	2:50	D
M.	10	2:40						2	4:50	F
	15	2:30					2	5	9:50	H
	20	2:30					4	15	21:50	J
	25	2:20				2	7	23	34:50	L
	30	2:20				4	13	26	45:50	M
	40	2:10			1	10	23	45	81:50	O
	50	2:10			5	18	23	61	109:50	Z
	60	2:00		2	15	22	37	74	152:50	Z
	70	2:00		8	17	19	51	86	183:50	Z
54	5	–						0	3:00	D
M.	10	2:50						3	6:00	F
	15	2:40					3	6	12:00	I
	20	2:30				1	5	17	26:00	K
	25	2:30				3	10	24	40:00	L
	30	2:30				6	17	27	53.00	N
	40	2:20			3	14	23	50	93.00	O
	50	2:10		2	9	19	30	65	128:00	Z
	60	2:10		5	16	19	44	81	168:00	Z
57	5	–						0	3:10	D
M.	10	2:50					1	3	7:10	G
	15	2:50					4	7	14:10	I
	20	2:40				2	6	20	31:10	K
	25	2:40				5	11	25	44:10	M
	30	2:30			1	8	19	32	63:10	N
	40	2:30			8	14	23	55	103:10	O
	50	2:20		4	13	22	33	72	147:10	Z
	60	2:20		10	17	19	50	84	183:10	Z

TABLE 2A

DEPTH mt.	NO DECOMPRESSION LIMITS	REPETITIVE GROUPS														
		A	B	C	D	E	F	G	H	I	J	K	L	M	N	O
3	–	60	120	210	300											
4.5	–	35	70	110	160	225	350									
6	–	25	50	75	100	135	180	240	325							
7.5	–	20	35	55	75	100	125	160	195	245	315					
9	–	15	30	45	60	75	95	120	145	170	205	250	310			
10.5	310	5	15	25	40	50	60	80	100	120	140	160	190	220	270	310
12	200	5	15	25	30	40	50	70	80	100	110	130	150	170	200	
15	100	–	10	15	25	30	40	50	60	70	80	90	100			
18	60	–	10	15	20	25	30	40	50	55	60					
21	50	–	5	10	15	20	30	35	40	45	50					
24	40	–	5	10	15	20	25	30	35	40						
27	30	–	5	10	12	15	20	25	30							
30	25	–	5	7	10	15	20	22	25							
33	20	–	–	5	10	13	15	20								
36	15	–	–	5	10	12	15									
39	10	–	–	5	8	10										
42	10	–	–	5	7	10										
45	5	–	–	5												
48	5	–	–	–	5											
51	5	–	–	–	5											
54	5	–	–	–	5											
57	5	–	–	–	5											

5 In order to calculate decompression requirements on repeat dives it is necessary to use tables 2A, 2B and 2C. These tables take into account the effects on the body of a relatively shallow first dive which would not have required decompression and is not therefore listed in table 1.

To calculate the repetitive dive group of the first dive, enter table 2A at the depth of the dive and select the duration of the dive. Next, move up the column to read the appropriate letter of the repetitive dive group. Incidentally, this table also lists "no decompression limits" as mentioned in note (1).

TABLE 2B

Surface Interval Credit Table

12:00 0:10	12:00 2:11	12:00 2:50	12:00 5:49	12:00 6:33	12:00 7:06	12:00 7:36	12:00 8:00	12:00 8:22	12:00 8:41	12:00 8:59	12:00 9:13	12:00 9:29	12:00 9:44	12:00 9:58	12:00 10:08
A	2:10 0:10	2:49 1:40	5:48 2:39	6:32 3:23	7:06 3:58	7:36 4:26	7:59 4:50	8:21 5:13	8:40 5:41	8:58 5:49	9:12 6:03	9:28 6:19	9:43 6:33	9:54 6:45	10:06 6:57
	B	1:39 0:10	2:38 1:10	3:22 1:58	3:57 2:29	4:25 2:59	4:49 3:21	5:12 3:44	5:40 4:03	5:48 4:20	6:02 4:38	6:18 4:50	6:32 5:04	6:44 5:17	6:56 5:28
		C	1:09 0:10	1:57 0:56	2:28 1:30	2:58 2:00	3:20 2:24	3:43 2:45	4:02 3:06	4:19 3:22	4:36 3:37	4:40 3:53	5:03 4:06	5:16 4:18	5:27 4:30
			D	0:54 0:10	1:29 0:46	1:59 1:16	2:23 1:42	2:44 2:03	3:04 2:21	3:21 2:39	3:36 2:54	3:52 3:09	4:04 3:23	4:17 3:34	4:29 3:46
				E	0:45 0:10	1:15 0:41	1:41 1:07	2:02 1:30	2.20 1:48	2:38 2:04	2:53 2:20	3:08 2:36	3:22 2:48	3:33 3:00	3:45 3:11
					F	0:40 0:10	1:06 0:37	1:29 1:00	1:47 1:20	2:03 1:36	2:19 1:50	2:34 2:06	2:47 2:19	2:59 2:30	3:10 2:43
						G	0:36 0:10	0:59 0:34	1:19 0:56	1:35 1:12	1:49 1:26	2:06 1:40	2:18 1:54	2:29 2:06	2:42 2:18
							H	0:33 0:10	0:54 0:32	1:11 0:50	1:26 1:05	1:39 1:19	1:53 1:31	2:04 1:44	2:17 1:56
								I	0:31 0:10	0:49 0:29	1:04 0:46	1:18 1:00	1:30 1:12	1:43 1:25	1:55 1:37
									J	0:28 0:10	0:46 0:27	0:59 0:43	1:11 0:56	1:24 1:06	1:36 1:19
										K	0:26 0:10	0:42 0:26	0:54 0:40	1:07 0:52	1:18 1:03
											L	0:25 0:10	0:39 0:25	0:51 0:37	1:02 0:49
												M	0:24 0:10	0:36 0:24	0:48 0:36
													N	0:23 0:10	0:34 0:23
														O	0:22 0:10
															Z

Follow up the line to the same letter in the "Surface terval Credit Table", 2B. This takes into account the time ent at the surface between dives. Move up the column to e level which covers time between dives. Next move along at column to the right to select the point of entry to table :. This will be in the form of a new letter.

TABLE 2C

DEPTH mt. — REPETITIVE DIVE DEPTH

mt.	12	15	18	21	24	27	30	33	36	39	42	45	48	51	54	57
A	7	6	5	4	4	3	3	3	3	3	2	2	2	2	2	2
B	17	13	11	9	8	7	7	6	6	6	5	5	4	4	4	4
C	25	21	17	15	13	11	10	10	9	8	7	7	6	6	6	6
D	37	29	24	20	18	16	14	13	12	11	10	9	9	8	8	8
E	49	38	30	26	23	20	18	16	15	13	12	12	11	10	10	10
F	61	47	36	31	28	24	22	20	18	16	15	14	13	13	12	11
G	73	56	44	37	32	29	26	24	21	19	18	17	16	15	14	13
H	87	66	52	43	38	33	30	27	25	22	20	19	18	17	16	15
I	101	76	61	50	43	38	34	31	28	25	23	22	20	19	18	17
J	116	87	70	57	48	43	38	34	32	28	26	24	23	22	20	19
K	138	99	79	64	54	47	43	38	35	31	29	27	26	24	22	21
L	161	111	88	72	61	53	48	42	39	35	32	30	28	26	25	24
M	187	124	97	80	68	58	52	47	43	38	35	32	31	29	27	26
N	213	142	107	87	73	64	57	51	46	40	38	35	33	31	29	28
O	241	160	117	96	80	70	62	55	50	44	40	38	36	34	31	30
Z	257	169	122	100	84	73	64	57	52	46	42	40	37	35	32	31

7 Table 2C is the "Residual Nitrogen Time" table. It provides information on what time must be added to the actual bottom time of the second dive in order to calculate the effective time of that dive for decompression calculations.

Enter the table on the left at the revised group letter and move along to the depth of the dive. The number shown in the box is the number of minutes to be added to the bottom time of the second dive. This is known as the "Penalty time" for a new dive at the selected depth.

For qualified medical practitioners only.

The following appendices are compiled from information given in a number of published references (listed in the bibliography) and from interviews with medical practitioners. The medical comments which follow are not intended as exhaustive instructions on treatment of marine-related accidents but should assist doctors who may not have such information immediately available to them. More detailed data is provided in the original texts and in various medical journals.

While every care has been taken to ensure that the medical information provided herein is correct, the publisher and author do not accept responsibility for complications arising from the use of this book.

The author would be interested to receive constructive comments upon the information provided, including case histories of marine bio-medical incidents occurring in the Red Sea region. It is hoped that future editions of this book may be updated as more information becomes available.

APPENDIX I

Medical Treatment of Puffer Fish Poisoning

1 During the early stages, prior to paralysis, a tube should be inserted for gastric lavage. Full anaesthetic facilities and suction equipment are essential to prevent the possibility of stomach contents being aspired into the bronchial tree. Emetics such as Apomorphine, 2-8mg s.c. or Ipecachuana syrup may be used.

2 Ideally, to control respiration, endotracheal intubation should be carried out. This will also prevent the possibility of aspiration of vomitus which is a real danger in cases of bulbar paralysis combined with gastrointestinal symptoms. If you are unable to carry out endotracheal intubation use whatever method is available to maintain respiration.

In relatively mild cases of respiratory impairment it may be enough to assist with intermittent positive pressure by, for example, use of a Bird respirator. If there is a rising concentration of CO_2 in arterial blood or an increasing respiratory rate there is a need for respiratory support but oxygen supplementation is not generally required.

In cases where respiratory depression is more severe and symptomatic distress or cyanosis occurs together with an increasing arterial CO_2 or a decreasing arterial O_2 it would be advisable to take control of respiration by endotracheal intubation and mechanical ventilation together with monitoring of O_2, CO_2 and Ph levels for about six hours. After this period the patient may be gradually weaned from the respirator over the next 12-24 hours.

3 It should be noted that fixed dilated pupils in an apnoeic and general unresponsive patient does not indicate "cerebral death" in cases of Puffer fish poisoning but is a common symptom associated with temporary paralysis of the patient.

4 At the time of writing there is no specific pharmacological therapy available. Vagal tone may be controlled by Atropine. 10% Calcium gluconate applied intravenously has been recommended as a non-specific stimulant which could improve the action potential of neurones. Once the paralysis starts to abate it may help to administer anticholinesterases (e.g. pyridostigmine or edrophonium) but these have no value during the acute phase of paralysis.

5 Fluid and electrolyte replacement is essential and medication should be administered intravenously. Vital signs such as serial haematocrit, S.G., electrolytes, C.V.P., urinary output and analyses should all be recorded. Vasodilation, which may occur in the initial stages may be countered by plasma expanders. In view of the possibility of haemorrhages occurring, cross-matched blood should be at hand.

6 Decisions regarding respiratory therapy should be based upon the results of serial monitoring of arterial blood gases and pH. E.K.G. and E.E.G. are indicated. If pulmonary oedema develops it may be necessary to increase inspiratory pressure to 60cm of water.

7 In view of the fact that the patient often remains conscious he should be continually reassured and a minor tranquilliser may be periodically administered. Medical discussions in front of the patient should be guarded.

8 In severe cases steroids such as hydrocortizone 200mg (repeated as indicated) could be beneficial.

9 Good nursing care is essential and particular attention should be given to pressure areas, eye and mouth toilets, etc.

APPENDIX 2

Medical Notes on Shark Liver Poisoning

The toxin is probably a parasympathomimetic substance.

1 First aid should commence with an emetic such as apomorphine 2-8mg s.c.

2 In order to control respiration which may become impaired the ideal method is endotracheal intubation which also serves the function of preventing aspiration of vomitus (a particular danger in cases of bulbar paralysis with gastrointestinal symptoms). Failing the possibility of endotracheal intubation, maintain respiration with any means available. See also notes for respiratory control in Puffer fish poisoning (Appendix 1) which also apply in this case.

3 Fluid and electrolyte replacement by intravenous means if indicated by monitoring (i.e. serial haematocrit, S.G., electrolytes, C.V.P., E.K.G., urinary output and analyses, etc.)

4 Treat convulsions as for status epilepticus.

5 Neuromuscular and Neurological features may be relieved by intra-venous administration of 10% Calcium gluconate.

6 Sedation should be achieved with non-respiratory depressants (e.g. diazepam 10mg i.v. repeated as required.). Opiates in small doses may be used to relieve pain.

7 Use of hydrocortisone (100mg i.v. every 6 hours) during the acute phase may be of value but specific pharmacological treatment is not presently available.

APPENDIX 3

Notes on Medical Treatment of Hydroid and Jellyfish Stings

a *Hydroids*

Application of local anaesthetic ointment is very effective in relieving pain. A cortico-steroid cream or ointment such as Ultralan 0.5% can be used to treat symptoms and reduce associated skin disorders. While methylated spirits or alcohol may be used as first-aid treatment they are less useful in the later medical stages.

Some patients may react more acutely than others and in severe cases cardiovascular and respiratory assistance may be necessary. In such cases of extreme distress tranquillisers and muscle relaxants have been recommended such as Diazepam 10mg by slow intravenous injection. Those allergy prone patients who have an exaggerated sensitivity to such stings may be assisted by i.v. hydrocortisone (for systemic reactions); s.c. adrenalin (for bronchospasms) or anti-histamines (for skin allergies).

It should be noted that while no deaths have been confirmed as a result of hydroid stings there have been a number of cases reported where hydroid stings were suspected to have been the cause of death.

b *Jellyfish stings*

Some of the most detailed clinical knowledge of jellyfish stings is to be found in Australia where Queensland doctors are regularly confronted with such cases. Several medical practitioners have written of their experiences in this field and further information may be obtained from these original sources.

Interested readers are referred to Cleland and Southcott (1965); Barnes (1960, 1963) and Edmonds (1974).

The following comments are a condensation of views expressed by these Australian researchers.

1 One of the most effective treatments for the pain of jellyfish stings is a local anaesthetic such as 5% Lignocaine.

2 Adrenalin (epinephrin) may be used after giving pain combating agents such as morphia.

3 Painful muscle spasms can be treated by 10ml Calcium gluconate i.v. or by 10mg of diazepam by slow. i.v.

4 Artificial respiration and oxygen are required in cases where depression of respiration occurs.

5 Cortisone or analogue therapy may be useful.

6 If severe stings occur on the limbs a tourniquet should be considered.

7 Anti-allergic therapy is sometimes indicated (e.g. hydrocortisone 100mg i.v. and repeat p.r.n. or in cases of respiratory allergy, adrenalin).

8 Once the initial symptoms have subsided subsequent treatment of local lesions may be carried out with anti-histaminic ointments; topical cortisone or with certain hand creams.

9 In cases where eyes have been stung non-aqueous local anaesthetic solution (e.g. 4% lignocaine eye drops or ointment followed by a steroid ointment) are suggested while all aqueous eye-drops should be avoided. Steroid antibiotic eye ointments such as Hydrophenicol or Sofradex have been successfully used in some cases.

APPENDIX 4

Medical Notes on Treatment of Cone Shell Stings

First aid treatment should include a ligature and incision of the wound in order to remove the venom as one would do for a snake bite. If signs of paralysis set in respiratory resuscitation may be necessary and in those cases which have been well reported it is the application of artificial respiration which has been regarded as being responsible for saving the victim's life. If circulatory depression occurs first ensure that the patient is warm and that the feet are elevated above the head and then if necessary apply external cardiac massage.

If medical facilities are available the following additional measures may be taken. In cases of respiratory paralysis endotracheal intubation is the best procedure since this will also prevent aspiration of vomitus and will assist with tracheobronchial toilet if this is indicated.

External cardiac massage etc. as indicated by E.K.G. monitor.

Special care should be given to pressure areas. Do *not* use any respiratory depressants.

APPENDIX 5

Medical Treatment of Stingray Vounds

tingrays can cause a deep wound into which hey may inject venom. The spine has recurved errations which render it difficult to extract. Deaths have occurred in cases where the spine as penetrated heart, abdominal or lung avities. There are however no known records of fatalities resulting from stings caused by the ed Sea Stingray – *Taeniura lymma.*

The main feature of stingray wounds is the evere pain which follows spine penetration. his may last for several days but the normal attern is for a build-up over the first two ours followed by a gradual improvement over he following 6-12 hours. Secondary infection quite common and this may cause additional ggravation of the wound over a longer period. he wound often gapes and bleeds profusely. he general area swells and the skin urrounding the wound generally turns bluish.

In addition to these basic features the ollowing symptoms have been reported on ictims of stingray wounds.

Osteomyeletis in underlying bone.
Diarrhoea, nausea, anorexia.
Tonic paralysis in affected limb or more generally.
Fainting, palpitations, hypotension, cardiac irregularities and ischaemia.
Breathing difficulties including pain on inspiration and coughing.
Fever.
Delirium.

Recommendations for medical treatment involve rapid action to alleviate the pain, to combat the effects of the venom and to prevent secondary infection. Pain can be treated by a local anaesthetic such as Lignocaine (without adrenalin) being infiltrated into the wound. Alternatively, a regional nerve block an be applied by local injections of Lignocaine (0.5% to 1.0%). More general pain relief can be given by systematic analgesics such as pethidine or codeine.

A U.S.A. doctor with experience of treating stingray wounds recommends 30mg of pentazocine lactate given intravenously and the same dose intramuscularly as a pain reliever. Demorol has also been found to be effective in controlling the pain.

If generalised symptoms appear to be developing further and are a cause of concern, a ligature may be applied.

The wound should be X-rayed to check for foreign bodies and these should be removed, if necessary, by surgery.

Broad spectrum antibiotics (e.g. tetracycline 250 mgm q.i.d.) together with a local antibiotic such as neomycin should be given as soon as possible.

As soon as the patient's condition permits, the wound should be thoroughly cleansed and, if necessary, sutured. All basic signs should be monitored and symptomatic treatment given for those clinical features which present.

Tetanus prophylaxis is standard since there has been a reported case of death from tetanus in a victim who suffered a stingray injury.

In view of the potential anaphylactic nature of the situation it has also been recommended that 40mg of methyl prednisolone sodium succinate is injected with decreasing follow-up on a five day schedule.

APPENDIX 6

Medical Treatment of Turkeyfish Stings

Turkeyfish venom is similar to that of other scorpionfish. It can cause hypotension, myocardial ischaemia, conduction defects, respiratory depression and neuromuscular abnormalities. It is soluble in water and is denatured by heat.

The general symptoms caused by Turkeyfish envenomations are described on p. 72. There are a few reported cases of deaths due to their stings and in many cases symptoms are severe.

Rapid first-aid essential (see p. 72) and this should be followed by bed rest and local anaesthesia using an anaesthetic (such as lignocaine 2%) which does *not* contain adrenalin. Recent experience in the U.S.A. also suggests the use of pentazocine lactate – 30mg i.v. and i.m. as a more general pain reliever. Debridement of the wound is necessary to prevent ulceration. A course of an antihistaminic compound, initially administered intramuscularly, is recommended. The situation is potentially anaphylactic and this is dealt with by initially injecting 40mg of methyl predinisolone sodium succinate with repeated injections of a gradually decreasing dosage – over a five day period. Tetanus prophylaxis should be given.

A local antibiotic such as neomycin applied at the commencement of medical treatment will help to prevent secondary infection occurring but if, despite such a precaution, there are signs of infection then an oral antibiotic such as tetracycline 250mg q.i.d. should be given.

Some doctors recommend use of Stonefish antivenom in serious cases. It is presently unclear how effective this is against Turkeyfish venom. There is also another antivenom which has been developed for Scorpionfish stings.

APPENDIX 7

Medical Treatment of Stonefish Stings

First Aid treatment is discussed on p. 78. Since the venom is unstable and heat labile, the use of a hot water bath or compress at an early stage is most helpful in denaturing any venom which remains near the surface of the skin at the site of the spine puncture.

The poison can also be broken down by a wide range of compounds. Emetine hydrochloride has an antagonistic action against the venom and it may be infiltrated into the wound in a solution containing 65 mgm/ml. Alternatively potassium permanganate (5% solution) and congo red have been recommended for early injection into the skin puncture.

If Stonefish antivenom is available the recommended instructions should be followed. In view of the fact that stonefish antivenom is made using horse serum and the consequent danger of horse serum allergic response by the patient, it is recommended that extreme care is taken when using the antivenom and that an initial trial of 10% of the recommended dosage is given to check on allergic sensitivity by the patient. The importance of this precautionary measure cannot be over-emphasised. If allergic symptoms do manifest themselves or if there is any evidence to suggest horse serum allergy do

NOT administer the antivenom. A Stonefish has 13 dorsal spines and each one contains 5 to 10mg of venom. 1ml of the antivenom neutralises 10mg of venom. The dosage therefore depends upon the number of spine punctures e.g. 1-2 punctures initially 1 ampoule 3-4 punctures (2 ampoules); 5-6 punctures (3 ampoules) and so on. The antivenom should be stored, away from light, at 2°C to 8°C and should not be frozen. It should be used immediately upon opening.

N.B. *Since the antivenom is prepared from horse serum it should not be given to people with horse serum allergy.*

Local anaesthesia – using an anaesthetic which does *not* contain adrenalin – should be infiltrated into and around the wound.

Incision and debridement of lesions, particularly if these occur on fingers or toes, is also recommended.

The major symptom of stonefish poisoning is extremely sharp and agonising pain. In cases which are left untreated abcess formation, necrosis and gangrene have occurred. The degree of shock varies in each case but may be quite severe. Muscle weakness and paralysis in the effected limb are frequently reported. Systemic effects are due to the myotoxin acting on all types of muscles including those of the respiratory and circulatory systems.

In addition to initial local treatment of the wound including cleaning and anaesthesia, and injection of the antivenom (if available); tetanus prophylaxis is prudent, together with a course of antibiotics to prevent secondary infection.

In severe cases cardiac massage and artificial respiration may be necessary and general monitoring of all body functions is recommended procedure. Care should be taken to avoid the clinical complications of bulbar paralysis.

Stonefish antivenom can be purchased from Commonwealth Serum Laboratories, 45 Poplar Road, Parkville, Victoria, Australia, 3052. Telephone: 389 1911; Telex AA 32789. The cost in 1985 was 28 Australian dollars per ampoule of 2000 units (i.e. sufficient for treatment of 1-2 punctures). Minimum order is A$250. Orders should be addressed to the Export Operations Manager. Delivery usually takes about 10 days.

APPENDIX 8

Treatment of Catfish Wounds

Catfish inflict a sting by means of one of three spines – a single dorsal spine, or one of two pectoral spines. The spines, which are located at the leading edge of the respective fins, are needle sharp, stout bones with backwardly curving teeth. They are mounted on swivel joints and can be voluntarily erected. Each spine is covered by skin which contain venom glands. The puncture wound may be lacerated as a result of withdrawal of the toothed spine.

Initial symptom of spinal puncture is extreme pain which radiates centrally. The wound is initially pale and then blue-purple and finally red and swollen. The swelling may become quite extensive but usually subsides within 48 hours. Bacterial infection of the wound frequently occurs.

In addition to the first aid measures described on p. 92, local anaesthesia will provide relief (use anaesthetic without adrenalin). The heat treatment which is fairly standard for marine fish envenomations has been experimented with in the case of catfish stings, and infra red heat has been succcessfully tested – administering the heat from an infra-red heat lamp for twenty minute sessions – repeated as necessary. General analgesics may be given (e.g. codeine), and local application of antibiotics such as neomycin is suggested. If secondary infection occurs systemic antibiotics are also indicated (e.g. tetracycline 250mg q.i.d.).

APPENDIX 9

Medical Notes on Treatment of Victims of Barracuda and Shark Attacks

At the site of attack

a Go immediately to the patient – do *not* ask an ambulance to bring him or her to you.
b Take transfusion equipment – materials in order of preference are Group O Rh negative blood, plasma, 5% albumen, haemacel, normal saline, dextran and other plasma expanders. Blood replacement levels depend upon clinical state measured by pulse blood pressure, cardiovascular pressure etc.
c Sedate the patient (e.g. Morphine 15mg i.v.)
d Set-up medical record so that this can be referred to at hospital.
e Only transfer patient to hospital after recovery from shock.

At Hospital

a Go straight to surgical operating theatre.
b Resuscitation first priority.
c When condition satisfactory mount full investigation including X-ray with radio opaque markers placed over site of possible bone damage.
d Surgical inspection including excision of necrotic material.
e Bacteriological culture.
f Conservative surgery advisable.
g Early skin grafting when possible.
h Tetanus prophylaxis.

There are usually secondary infection complications with shark wounds and a paracolon bacillus resistant to penicillin has been isolated as has *Clostridium tetanus*.

9A

SHARK ATTACKS

Characteristics of shark wounds.

Often extensive wound with a ragged edge. Considerable tissue damage and bleeding. Shark teeth and other fragments may remain in wound. Skin abrasions may also occur due to shark's rough skin. Death rate is proportional to the number of severed blood vessels.

Grade 1: Fatal
Both femoral arteries
One femoral + one posterior tibial artery
One femoral artery in upper 1/3rd of thigh

Grade 2: Survive if good First Aid
One femoral artery in lower 2/3rds of thigh
One brachial artery
Two posterior tibial arteries
Abdominal wounds with bowel involvement

Grade 3: Always survive if properly treated
One posterior tibial artery
Superficial limb wounds; no vessels cut
Superficial abdominal wounds; no peritoneal involvement.

B

ARRACUDA ATTACKS

ledical attention, following first aid requires hat adequate haemostasis is established. Only vhen the clinical state is stable should the atient be moved. Blood loss should be eplaced with intravenous fluids as determined y monitoring pulse, blood pressure and ardiovascular pressure.

Sedatives such as Diazepam by slow i.v. or orphine may be given and one should be eady to give artificial respiration. Once ospital facilities are available bacteriological ulture of the wounded tissues should be ndertaken. An X-ray may be indicated, and e wound will require a blood transfusion so ross matching should be carried out and itable blood held nearby. The wound may be eated with a topical antibiotic such as eomycin.

Tetanus prophylaxis is standard procedure, d a broad spectrum antibiotic may also be nsidered a wise precautionary measure.

PPENDIX 10

ledical Treatment of Moray Eel ites

nere has been considerable debate over the st century regarding whether or not Moray els secrete a venom which is injected into eir victims when they bite. Until recently the nsensus was that there is no venom apparatus volved or that venom may be secreted in the latine mucosa and introduced into the ound by a straightforward process of contact and cross-injection. There have been several quite serious incidents in recent years which have lent credence to the belief that toxin is involved in bites from large Indo-Pacific Morays.

Moray wounds are often torn and ragged. Profuse bleeding is usually associated with Moray bites and secondary infection is common. The victim usually becomes pale, exhibits a rapid pulse hypotension and may faint on standing.

Medical attention should be directed at arresting bleeding and establishing adequate haemostasis. The patient is best left unmoved until the condition stabilises. Lost blood should be replaced with intravenous fluids as indicated by monitoring of body functions.

Sedatives can be given but one should remain ready to take over respiration. Once the patient has been transported to hospital bacteriological culture and radiological assessment should be undertaken. The wound should be surgically cleaned and repaired. A topical antibiotic such as neomycin can be used to help with ensuring adequate antisepsis of the wound.

Tetanus prophylaxis is standard procedure and a broad spectrum antibiotic may also be considered prudent.

Acknowledgements

The preparation of this book would not have been possible without the help of many scientists who have studied potentially harmful Red Sea creatures. Their work is referenced in the text and I gratefully acknowledge the vital contribution which such studies have made to our knowledge of Red Sea marine life. Not all the relevant investigations have taken place in the Red Sea but may have been carried out on species whose distribution spreads from the Red Sea through the Indian Ocean and across at least part of the tropical Pacific. Where such studies are of direct relevance to the Red Sea environment their results have been incorporated into the text. In a global review of the subjects of venomous invertebrates and dangerous sharks, special mention must be made of the contributions made by Australian divers, marine scientists and medical practitioners. While we are fortunate the most dangerous of Australia's venomous marine species do not occur in the Red Sea (e.g. box jellyfish, sea-snakes and blue-ringed octopus), we must nevertheless acknowledge the considerable strides which have been made in Australia towards treatment of a wide range of coral-reef associated injuries. The most notable is that developed for dealing with Stonefish stings and comprises an antivenom which is manufactured by the Commonwealth Serum Laboratories in Melbourne, Australia. This antivenom has been used to relieve the pain and suffering of several people who have stepped on Stonefish in the Red Sea.

On a more personal note I wish to thank the friends who have dived with me over the years in the Red Sea. In particular I should like to acknowledge the help and encouragement of the following: Dr. J.E. Randall, Dr. R.F.G. Ormond, Richard Moore, Douglas Allen, Dr. Dirar Hassan Nasr, Captain Abdel Halim, Captain Abdel Nebi, Hagen Schmid, Georg Jungbauer and Robin Lehman. In addition to these diving partners I also wish to gratefully acknowledge the help which I have received from the University of Khartoum, Professor Yousif Abu Gideiri, the Marine Fisheries Research Centre at Jeddah, Saudi Arabia, The Oceanography Institute, Port Sudan, and from the Cambridg Marine Research Group which established a laboratory in Port Sudan in the 1970's.

Preparation of the book has been greatly aided by a number of highly skilled underwater photographers who have contributed their material to this publication. They are separately acknowledged by name and I wish to add my personal thanks to them for their generosity and cooperation. I am also indebted to illustrator Jane Starck whose excellent drawings and paintings help to illuminate the text.

Medical practice and pharmacology are constantly changing fields where last year's advice can be easily outdated by more recent research or new drugs. For this reason the medical information is of a general nature and recommended medicines are widely used ones I am grateful to the following medical practitioners who have checked and made useful comments on medical aspects of the book: Dr. C. Casey, Dr. John Casey. Finally I want to express my sincere thanks to Mme. Rafi Minhas who did such an excellent job of typing the text.

REFERENCES

dachi, R. and Y. Fukuyo (1979)
The thecal structure of a marine dinoflagellate *Gambierdiscus toxicus* gen. et sp. nov. collected in a ciguatera-endemic area. *Bull. Jpn. Soc. Sci. Fish. 45* 67-71.

nner, A.M. and J.E. Randall (1952)
Preliminary report on marine biology study of Onotoa atoll, Gilbert Islands. *Atoll Res. Bull. 13*: 43-62.

rnes, J.M. (1960)
Observations on jellyfish stingings in North Queensland.
Med. J. Aust. 2: 993-999.

rnes, J.M.(1963)
Stingings by jellyfish,
pp. 103-105, In: International Convention on Life Saving Techniques, Sydney, March 11th-20th, 1960. "B" Group – Scientific Section. "The hazards of dangerous marine creatures", published by the Committee of the 1960 International Convention of Life Saving Techniques, February, 1963, pp. 1-131.

rnes, J. (1964)
A dangerous starfish – *Acanthaster planci* (Linne)
The Med. J. of Australia, April 18, 1964. pp. 592-593.

eland, J.B. and Southcott, R.V. (1965)
"Injuries to Man from Marine Invertebrates in the Australian Region". National Health and Medical Research Council, Department of Health, Commonwealth of Australia, Special Rep. Ser. No...12.

Cooper, M.J. (1964)
Ciguatera and other marine poisoning in the Gilbert Islands.
Pac. Sci. XVIII Oct. 411-440.

Edmonds, C. (1974)
Dangerous marine animals of the Indo-Pacific Region.
Australian Sports Publications, Melbourne, pp. 235.

Flecker, M. (1936)
Cone shell poisoning, with report of a fatal case.
Med. J. Aust. 1:464-466.

Gohar, H.A.F. and Mazhar F.M. (1964)
The elasmobranchs of the north western Red Sea.
Publ. Mar. Stn. Al-Ghardaga (Red Sea) 13: 1-144.

Halstead, B.W. (1965)
Poisonous and venomous marine animals of the world. Vol. 1. U.S.Govt. Printing Office.

Halstead, B.W.
Poisonous and venomous marine animals of the World, Vols. 2 and 3. U.S. Govt. Printing Office.

Khlentzos, C.T. (1950)
Seventeen cases of poisoning due to ingestion of an eel, *Gymnothorax flavimarginatus. Am. J. Trop. Med., 30*:785-793.

Kohn, A.J. (1958)
Cone shell stings. Recent cases of human injury due to venomous marine snails of the genus *Conus. Hawaiian Med. J., 17* (6): 528-532.

Neve, P. (1972)
Dangerous Red Sea Fishes. *Bull. Mar. Res. Centre, Saudi Arabia,* n 1,1-12.

Pickwell, G.V. and Evans, W.E. (1972)
Handbook of Dangerous Animals for Field Personnel, Naval Undersea Centre, San Diego, California.

Randall, J.E. (1958)
A review of ciguatera, tropical fish poisoning, with a tentative explanation for its cause. *Bull. Mar. Sci. Gulf and Carib. 8(3):236-267.*

Randall, J.E. (1969)
How dangerous is the Moray Eel? *Aust. Nat. Hist.* 16(6):177-182.

Randall, J.E. and Helfman, G.S. (1973)
Attacks on humans by the Blacktip Reef Shark (*Carcharhinus melanopterus*). *Pac. Sci., 27*:226-238.

Randall, J.E. (1977)
Contribution to the biology of the Whitetip Reef Shark (*Triaenodon obesus*). *Pac. Sci., 31*:143-164.

Randall, J.E. (1980)
A survey of ciguatera at Enewetak and Bikini, Marshall Islands, with notes on the systematics and food habits of ciguatoxic fishes. *Fishery Bulletin 78*(2):201-249.

Taylor, F.J.R. (1979)
A description of the benthic dinoflagellate associated with ciguatoxin including observations on Hawaiian material. In: D.L Taylor and H.M. Seliger (editors) Toxic dinoflagellate blooms, pp. 71-76, Elsevier, Holland.

Taylor, R. and Taylor, V. (1981)
The great Shark Suit Experiment. 64pp. published by
Ron Taylor Film Productions Pty. Ltd. Australia.

Taylor, V. (1981)
A jawbreaker for sharks. *Nat. Geog. 159*(5):664-667.

INDEX